Vision

How It Works and What Can Go Wrong

John E. Dowling and Joseph L. Dowling, Jr.

D1481510

The MIT Press
Cambridge, Massachusetts
London, England

SOUTH COUNTRY LIBRARY
22 STATION ROAD
BELL~~~~ ~~~~713

MAY 23 2016

© 2016 Massachusetts Institute of Technology

All rights reserved. No part of this book may be reproduced in any form by any electronic or mechanical means (including photocopying, recording, or information storage and retrieval) without permission in writing from the publisher.

This book was set in Stone Serif and Stone Sans by Toppan Best-set Premedia Limited. Printed and bound in the United States of America.

Library of Congress Cataloging-in-Publication Data

Names: Dowling, John E. | Dowling, Joseph L., Jr.
Title: Vision : how it works and what can go wrong / John E. Dowling and Joseph L. Dowling, Jr.
Description: Cambridge, MA : The MIT Press, 2016. | Includes bibliographical references and index.
Identifiers: LCCN 2015039901 | ISBN 9780262034616 (hardcover : alk. paper)
Subjects: LCSH: Vision—Popular works. | Eye—Physiology—Popular works. | Ophthalmology—Popular works. | Eye—Diseases—Popular works.
Classification: LCC QP475.5 .D69 2016 | DDC 612.8/4—dc23 LC record available at http://lccn.loc.gov/2015039901

10 9 8 7 6 5 4 3 2 1

In memory of our father
Joseph L. Dowling, Sr., MD
Who introduced us to eyes and was the first to specialize in
ophthalmology in Rhode Island

Contents

Preface

This book is about vision—what we currently know about how it works, how we perceive things, and what can go wrong and compromise our sight. The visual system is arguably our best understood sensory system, and vision is our major sensory system. About 40% of our sensory input is visual, and about 50% of our cerebral cortex is devoted to processing visual information. By comparison only about 13% of the cortex is involved in analyzing auditory information.

For most animals vision is paramount, but some species use other senses to explore the world—ants and mice use smell as their primary sense, and whales and bats use hearing. It is also true that various animals use their visual systems for different purposes. We have an especially high acuity, which enables us to see fine print, watch television, drive, and carry out other tasks requiring an ability to discriminate small objects. Many fish and reptiles have elaborate color vision mechanisms, more elaborate than we have, and they use color vision to recognize other members of their own species, members of the opposite sex, and even age—juvenile fish may be colored differently from adults. Other species, such as frogs and rabbits, are very responsive to movement, and this is reflected in the makeup of their visual systems. Every visual system has common features, and knowledge of any one has implications for all, including our own. In this book, we focus on vision in mammals, especially humans, and describe what can go wrong in the visual system. Both normal and defective visual systems shed light on how we see.

We have spent our professional lives researching the basic mechanisms of vision (John) and studying and treating visual diseases (Joe). We are often asked how we became interested in vision and ophthalmology. Although we did hear a bit about eyes at home while growing up (our father was an ophthalmologist), our careers evolved from different experiences as detailed in the section about the authors that follows.

During the past half-century, while we have been studying visual mechanisms or treating visual disorders, enormous progress has been made on both fronts. However, the basic scientist is often not aware of the progress being made on the clinical front and the puzzles that remain, and the clinician is not always cognizant of the more recent basic science discoveries. In this book we cover and relate basic information concerning visual mechanisms with various clinical abnormalities. We do this not only for the basic scientist or clinical specialist but also for a more general audience. As science and medicine become more and more specialized, it seems to us worthwhile to take a step back and see where we are. What do we know about how we see, and what is the current status of the major eye diseases? What causes them, and how can we treat them? And what does the future hold?

The great majority of visual problems originate in the eye, and we put most emphasis there, starting with how the cornea and lens project an image onto the light-sensitive elements inside the eye, the photoreceptors, and how that process can be compromised. We then discuss how the distal photoreceptor cells capture photons of light and how this primary step in vision can go awry and degrade vision. But the eye contains more than an array of photosensitive units; indeed, the first steps in the analysis of a visual image occur in the retina, which consists of many kinds of neurons that form a complex neural network. We devote much attention to how this network processes visual signals and describe two major diseases that affect retinal function.

From the eye, visual information travels to higher visual centers, and it is in the cortex that we perceive objects. Abnormalities in the cortical neural networks compromise vision in interesting and surprising ways, and a number of these defects are discussed.

The final chapter provides a brief overview of the unfolding of our knowledge of how vision works, from the Greeks to the present day. But what does the future hold? Many causes of eye disease and blindness can currently be prevented or cured, but there are still many blinding diseases for which we have no cure. Will we ever be able to cure blindness or provide a substitute for vision for those afflicted with an incurable blinding disease? We end with some speculations, but the main message of these stories is the recent and exponential increase in our knowledge of how we see, what may cause the major visual diseases, and how we might someday cure these severe and blinding disorders.

About the Authors

John

The person who influenced me most to study vision was a professor at Harvard. I was an undergraduate premedical student when I took a biochemistry course in my junior year with George Wald. The first semester of the course described the great metabolic pathways in organisms—glycolysis, respiration, and photosynthesis—that result in the formation of energy-rich molecules that all cells need to carry out their many tasks. There is much to memorize in learning these pathways, and many students find this biochemistry tedious. But Wald was mesmerizing in his presentations. He made the material come alive and a pleasure to learn. The second semester of the course, an elective, called "Topics in Biochemistry," was for me especially exciting. Wald's lecture on Albert Szent-Györgyi's research on muscle remains with me to this day. Szent-Györgyi, who won a Nobel Prize for his discovery of vitamin C, showed that after a piece of muscle was extracted with glycerol, leaving only the proteins behind, it would contract again if presented with an energy-rich molecule, adenosine triphosphate (ATP). To me this was phenomenal—getting at the essence of life.

Subsequently, Wald described his own work on the nature of light-sensitive molecules found in the photoreceptors in the eye. Wald had found that a derivative of vitamin A, vitamin A aldehyde now called retinal, combines with a protein to form the photosensitive visual pigments. Indeed it is retinal that absorbs the light and initiates visual excitation. This work, for which he was awarded a Nobel Prize, fascinated me. I asked Professor Wald if I could do research in his laboratory in the summer between my junior and senior years, and he agreed. The summer was an eye-opener (pun intended), and I continued to work in Wald's lab all during my senior year and the next summer also. Although I hadn't realized it at the time, I was hooked!

I headed off to medical school that fall but on free afternoons returned to Cambridge to continue research in the lab. I worked in the lab the following summer and throughout my second year of medical school. At that point I decided to take a leave of absence from the medical school—to return to the lab for one last year. However, I am still on that leave of absence—some 57 years later!

Under Wald's mentorship, I obtained a PhD and stayed on at Harvard for 3 more years, when I was recruited to the Wilmer Institute of Ophthalmology at Johns Hopkins University. I stayed for 7 years and then returned to Harvard, where I have been since. In Wald's laboratory, I first studied photoreceptors—what happens to them when an animal is deprived of vitamin A and, in my own lab, what happens to photoreceptors when they are exposed to dim and bright lights. I also studied a rat that had an inherited retinal defect that resembles the human disease retinitis pigmentosa.

Shortly before I moved to Hopkins, I made an observation that fascinated me—namely that a type of retinal cell thought to be a nonneural supportive (glial) cell contacted photoreceptor synaptic terminals. The questions then became to which cells do the photoreceptors talk, and what are the connections of cells downstream of the photoreceptors? These questions occupied me and my colleagues at Hopkins for the next several years. Once we had some inkling of the wiring of the retina, we next studied how the retinal cells respond to light. This was a long-time focus of my lab both at Hopkins and at Harvard, followed by a study of the mechanisms by which one cell communicates with another cell and, finally, how the retina develops. Most recently, we have focused on color vision mechanisms using zebrafish, which have an extensive color vision system.

Joe

I attended college at Brown University during World War II as a member of the Navy's V-12 program for premedical students. My interest in science and medicine was reinforced and amplified by the renowned bacteriologist Charles A. Stuart, with whom I took several didactic courses and a senior research seminar. My first research project was a study another student and I undertook on cows to identify the microorganisms present in various parts of the gastrointestinal tract. Professor Stuart had arranged for us to visit a local slaughterhouse, where we took our cultures. I don't recall the specific results of that study but it did introduce me to the intricacy and complexity of biological systems and research protocols. And I did get a mention in the published article.

After Brown, I matriculated to medical school, and there I decided to pursue a career in ophthalmology. Of course I was familiar with ophthalmology from my father, who was the first ophthalmologist in Rhode Island (previously there had been only combined eye, ear, nose, and throat specialists). I chose ophthalmology because it encompasses essentially all aspects of medical practice including medical diagnosis and treatment, surgery, emergencies, preventive medicine, chronic conditions, and

ongoing family care, similar to general practice. After an intern-
ship, I was called to active Naval duty during the Korean War. I
served as the medical officer on a ship assigned to the Sixth Fleet
in the Mediterranean.

My first experience in ophthalmology, with which I had had
no training or knowledge, occurred one morning as our ship was
proceeding across the Atlantic Ocean. It was an unexpected and
dramatic introduction to traumatic eye pathology. A crew member had sustained an eye injury when he cut the wire on a crate
of oranges and the wire snapped up and struck his face and eye.
Arriving in sick bay I observed his eye had a distorted pupil, a
complete laceration of the eyeball with the iris protruding from
the wound. I retreated to my tiny office where the Navy had
provided a medical book titled *The Specialties in General Practice.* I located the section on ophthalmology and looked up "iris
prolapse." The short and curt reference stated "This is a serious
emergency and the patient should be immediately referred to an
ophthalmologist." It also warned that under no circumstances
should an attempt be made to replace the iris. Since the near-
est ophthalmologist was 700 sea miles away, the challenge was
all mine. In our little operating room the eye was locally anes-
thetized. With the smallest surgical instruments we had, which
were more suitable for a whale's eye than a human one, I gen-
tly grasped the protruding iris and cut it off. I was gratified to
see the remains of the iris retract, causing the pupil to resume
an almost normal round shape. I sutured the incision with the
finest suture material available, which was about 10 times the
diameter of the largest suture material we would currently use in
eye surgery. Fortunately, the patient did well. The wound healed
well, he regained vision, and when we arrived in Europe, he was
flown back to a military hospital in the United States. Thus, my
surgical career in ophthalmology had begun.

After I was discharged from the Navy, I worked for several months assisting on research projects in the Howe Research Laboratory at the Massachusetts Eye and Ear Infirmary prior to beginning my residency there.

My 3 years in residency were a wonderful experience. My mentors were many of the outstanding ophthalmologists in the country and the world including Paul Chandler, Morton Grant, and Charles Schepens, to name just a few. I had a particularly rewarding experience with Taylor Smith, Chief Ophthalmic Pathologist and renowned retina surgeon. We published my first paper in the ophthalmic literature and won a prize for the best original paper presented at the American Medical Association Section Meeting on Ophthalmology in 1956. The paper was on the ocular manifestations of a rare disease known as Takayasu or pulseless disease. The major importance of the paper was on a technique we had devised for studying the retinal blood vessels by filling them with India ink. This technique and related techniques have been used extensively since (see figure 4.13) in studying retinal blood vessels in a variety of retinal diseases. Our paper was subsequently included in the book titled *Classic Papers in Ophthalmic Pathology.*

After finishing my training at the Massachusetts Eye and Ear Infirmary, I entered private practice in Providence, Rhode Island, where I have been happily occupied for the past 59 years. In 1991 I founded the Rhode Island Eye Institute, a group practice that now has 16 physicians specializing in all aspects of ophthalmology. Over the years I have published several papers in the ophthalmic literature and two textbook chapters. I have also served as an Associate Professor of Ophthalmology at the Brown University Medical School, where I am now emeritus.

Acknowledgments

We owe many thanks to the reviewers of the book, especially Len Levin, who read an early version and made superb suggestions regarding its organization. Jack Werner and two anonymous reviewers of the submitted manuscript also made substantial suggestions, virtually all of which we incorporated. Clearly, the book was greatly improved by the reviewers, and we greatly appreciated their time and effort.

Bob Prior of the MIT Press believed in the book and shepherded it through to publication along with his colleagues, especially Chris Eyer and Katherine Almeida. Renate Hellmiss and her team at mcb graphics expertly rendered the new figures and prepared all of the illustrations for publication. Kendal Kelly entered countless versions of the manuscript into the computer and carried out the numerous tasks such a project entails. We also thank Mark A. Hamel, Cecile M. Lacourse, Lory S. McCoy, and Robert L. Bahr of the Rhode Island Eye Institute for providing the clinical illustrations.

The first version of the manuscript was written during a delightful 2 months J.E.D. spent at the Hanse-Wissenschaftskolleg in Oldenburg, Germany. Many thanks to its Director, Reto Weiler, and his colleagues, who made that visit so memorable. J.E.D.'s research has been generously supported over the years by the

National Institutes of Health and the National Eye Institute. We gratefully acknowledge funds from the Department of Molecular and Cellular Biology and the Faculty of Arts and Sciences, Harvard University for preparation of the figures and other publication costs.

1 Visual Pathways, Eye Development, and Retinal Organization:

Visual Deficits and Blindness

Virtually everyone knows that Beethoven was deaf when he composed his greatest works. Fewer know that John Milton, one of England's greatest poets, was blind when he wrote his greatest epic poems, Paradise Lost, Paradise Regained, *and* Samson Agonistes. *Born in 1608, Milton became totally blind when he was 43. Perhaps his most famous sonnet, XIX, first published in 1673, a year before his death but probably written shortly after he became totally blind, reflected on his blindness. In its opening eight lines (below), Milton expresses despair that his affliction prevents him from using his great talent to glorify God. Should God expect him, a blind man, to produce great works? But, this Milton did accomplish, by dictating his epics, line by line, to his three daughters. Why Milton went blind is unknown.*

Sonnet XIX

When I consider how my light is spent,
E're half my days, in this dark world and wide,
And that one Talent which is death to hide,
Lodg'd with me useless, though my Soul more bent
To serve therewith my Maker, and present
My true account, lest he returning chide,
'Doth God exact day-labour, light deny'd?'
I fondly ask ...

It is through our sensory systems that we are in contact with the outside world. Without them we would be isolated not only from the environment but from each other. We have five senses—sight, hearing, touch, smell, and taste. Some animals have additional sensory systems. For example, certain fish can detect electrical fields; birds, magnetic fields; and snakes have specialized heat sensors.

Surveys consistently report that people fear total blindness more than any other disability. The major risk factor for eye disease is age. Those of age 80 or older make up less than 10% of the population (although that percentage is growing), but 69% of the blindness in the country occurs in those over age 80. But many people under age 80 are afflicted with sight-impairing diseases. The most common age-related eye diseases are cataract, macular degeneration, glaucoma, and diabetic retinopathy. Visually impaired individuals say they would trade away many of their remaining years to regain good vision. Indeed, legally blind patients say they would be willing to trade 36% of their remaining years to regain perfect sight, and those whose vision is limited to only light perception indicated they would trade 74% of their remaining years to see again. And, tragically, every 7 minutes on average someone goes blind in the United States from one cause or another.

Deafness is also isolating, and when blindness and deafness occur simultaneously, it is even more devastating, as happened to Helen Keller at a young age. She was alone, depressed, and quite angry until her dedicated teacher, Annie Sullivan, made contact with her through the sense of touch. The education of Helen Keller is the familiar story of how Annie Sullivan communicated with her, introduced Helen to her environment, to other human beings, and very much more. Together, Helen and Anne achieved remarkable things, including Helen's earning a college degree from Harvard's sister school, Radcliffe, in 1904, the first blind and deaf person ever to earn a college degree!

Much of what Helen labored to discover about the world we learn easily and naturally with our intact sensory systems. Before Annie arrived, Helen had been enormously frustrated by her inability to communicate and understand the world. The breakthrough came one day when Annie and Helen were in the family pump house. As water from the spout rushed over one of Helen's hands, Annie spelled w-a-t-e-r in the palm of her other hand. The sensation of the water rushing over one hand while simultaneously feeling the letters spelled out in the other was transforming for Helen: she instantly understood and wanted to know the names of everything she touched. Within a few hours she had learned the names of 30 objects and began her understanding of the world.

Vision is dominant of our five senses, partially because it is remarkably adaptable. For example, when an eye has been in the dark for half an hour or so, it "dark adapts." The retinal rod photoreceptors become exquisitely sensitive, capable of responding to just a single photon. When the eye is exposed to light, the cone photoreceptors come into play, and we see details and color. Visual acuity in the central (foveal) region of the retina is extremely acute, allowing us to distinguish the tiniest of objects. Even so, certain of the capabilities of the human visual system are not as extraordinary as those of some animals. Many fish, reptiles, and birds have far richer color vision systems than do humans and can perceive wavelengths of light that the human eye cannot see. And certain birds have two foveal regions—one to see details in front and a second to see details from the side, such as a small mouse running below on the ground.

Visual Pathways

Vision occurs when rays of light are captured by the photoreceptor cells of the retina that line the back of the eye. This requires a
transparent window into the eye and a mechanism for bringing
the rays of light to a focus on the photoreceptors. Two remarkable structures, the cornea and lens, serve to accomplish this
(figure 1.1). The eye also has a rigid and spherical shape, and
how this happens is now quite well understood.

Lining the back of the eye and making up the retina are not
only the photoreceptors but a neural network, which is a true
part of the brain pushed out into the eye during development.
Consisting of five major classes of neurons, the retina performs
the initial stages of analyzing form, color, and movement of an
image (figure 1.1). This information is encoded in the output
neurons of the retina, the ganglion cells, which transmit all
visual information to the rest of the brain.

From the retina, most of the visual message in mammals
goes first to a brain structure called the thalamus, specifically
to a highly organized cluster of cells called the lateral geniculate
nucleus (LGN). (Note that a brain nucleus is a cluster of neurons concerned with a specific function or task.) Neurons in the
LGN, in turn, transmit visual information to the cortex, where
features of the visual image are analyzed further and in exquisite
detail (figure 1.2). In non-mammalian species most of the visual
information from the retina goes to a large brain area called the
tectum. In nonmammals the cerebral cortex is a primitive brain
structure concerned largely with analyzing smell, and the tectum is the major area involved in analyzing sensory information
and initiating behavioral responses. It is the dramatic evolution
of the cerebral cortex into a structure involved with processing
and integrating all sensory information, the initiation of movements, memory and learning, and so forth that characterizes

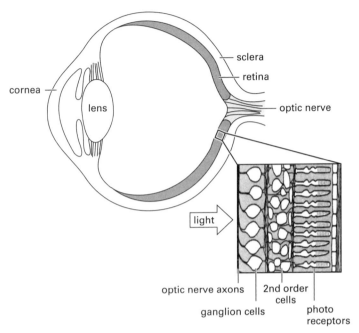

Figure 1.1
The retina, consisting of photoreceptors, several kinds of second-order cells, and ganglion cells, lines the back (sclera) of the eye. The axons of the third-order ganglion cells run along the surface of the retina and collect together to form the optic nerve that exits the eye. Light enters the eye via the transparent cornea and is focused by the cornea and lens onto the photoreceptors.

mammalian brains and accounts for the development of higher brain function, reaching its pinnacle in humans.

In the cortex all visual information is first analyzed in an area in the posterior part of the brain called visual area 1 (V1). From there it progresses to area V2, and then to a variety of areas—V4, V5, and so on—that are specialized to analyze one or another aspect of the visual image. Beyond these areas are

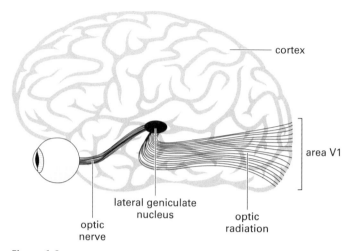

cortex

area V1

lateral geniculate
nucleus

optic
nerve

optic
radiation

Figure 1.2
The central visual pathways. The optic nerve extends from the eye to the
lateral geniculate nucleus (LGN) in the thalamus. Neurons from the LGN
extend their axons via the optic radiation to the primary visual area (V1)
at the back of the cortex.

many more specialized areas—perhaps as many as 30–40—that
are concerned with even more specific aspects of visual perception, including, for example, object and face recognition. Figure 1.3 shows the basic flow of visual information from area V1
to these other cortical areas. From V1, most visual information
goes to adjacent V2; then different modalities—form, color, and
motion—go on to V4, V5, and so forth. From these areas there
are two basic visual pathways. One, termed the "what" pathway by Mortimer Mishkin and colleagues, is involved mainly
with object recognition; the other, the "where" pathway, is concerned with observing an object in space. All of this is described
in detail in chapter 6. The rest of this chapter is concerned with
the origins and organization of the eye in general and the retina

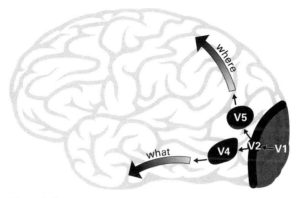

Figure 1.3
Visual areas beyond V1. Visual information goes from area V1 to area V2 and then to several separate areas including V4 and V5. Visual information progresses to higher visual areas via two pathways—a dorsal pathway called the "where" pathway and a ventral pathway termed the "what" pathway.

in particular. At the end of the chapter, we discuss various visual deficits, where they arise, and their impact on vision.

Development of the Brain and Eye

The nervous system of all vertebrates begins to develop in the embryo from a small group of cells on the dorsal (top) side of the embryo, called the neural plate. In humans the neural plate forms about 3 weeks after conception and initially consists of about 125,000 cells (figure 1.4). The neural plate cells derive from an outer layer of cells called ectoderm, which is destined to become mainly skin. There are two other layers that make up the early embryo: the mesoderm, which develops into muscle and bone, and the endoderm, which becomes the internal organs including the stomach and intestines.

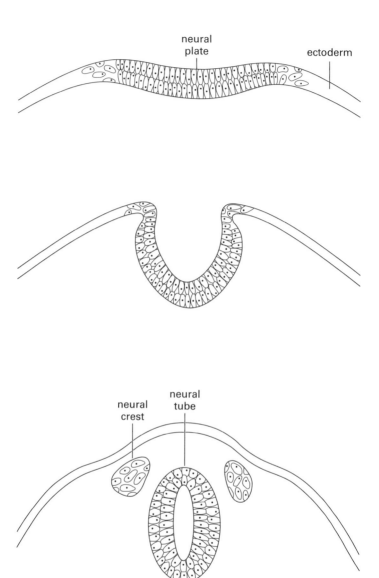

Figure 1.4

Formation of the neural tube and neural crest cells from the neural plate. During the third and fourth weeks of development of the human embryo, the neural plate invaginates, forming the neural tube. The neural tube differentiates into the central nervous system—the brain and spinal cord. The neural crest cells are derived from neural plate cells positioned laterally along the plate and left behind during the formation of the neural tube. Neural crest cells differentiate into the peripheral nervous system.

Between the third and fourth weeks of development in the human embryo, the neural plate folds inward to form a deep groove that eventually closes into a long tube, the neural tube, which separates from the overlying cells (figure 1.4). All of the brain derives from the neural tube; the anterior end becomes the brain proper, whereas the posterior part becomes the spinal cord. Some cells along the margin of the neural plate separate off to become the neural crest. The neural crest cells develop mainly into the peripheral nervous system, neural tissue outside of the brain and spinal cord.

The eye and retina derive from the anterior part of the neural tube, which evaginates to form two optic vesicles (figure 1.5A). Each optic vesicle subsequently invaginates to form an optic cup. The cells on the inner wall of the optic cup proliferate, leading to the formation of what is called the neural retina, whereas the cells on the outer wall of the optic cup become the pigment epithelium, a single layer of cells just distal to the photoreceptors (figure 1.5C). The photoreceptor cells form along the outer side of the retina. The photoreceptors and pigment epithelium, adjacent to each other, interact extensively throughout life. Indeed, if the retina separates from the pigment epithelium—a retinal detachment—as the result of a blow to the head or other trauma to the eye, vision is lost in the detached part of the retina. The reasons vision is lost when the retina is separated from the pigment epithelium are twofold; first, because the photoreceptors require the pigment epithelium to function properly (chapter 3), and, secondly, most of the nutrients and oxygen required for the functioning of the outer retina come through the pigment epithelium from a blood supply found overlying the pigment epithelium, a vascular bed of tissue called the choroid. Unless measures are taken promptly to reattach the retina to the pigment epithelium, the photoreceptors will degenerate in the detached part of the retina, and vision will be permanently lost.

A neural tube

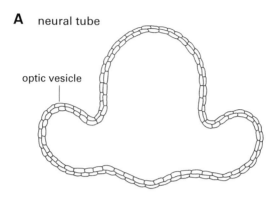

optic vesicle

B optic cup

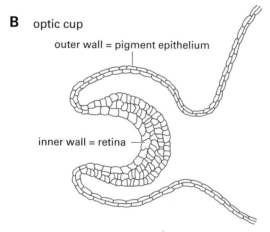

outer wall = pigment epithelium

inner wall = retina

C retina

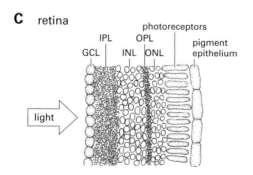

photoreceptors

IPL   OPL

GCL   INL   ONL   pigment epithelium

light

Figure 1.5
Schematic diagrams of the development of the retina from the neural tube in the embryo. (A) In the head region of the embryo, optic vesicles form from outpocketings of the neural tube. (B) The optic vesicles invaginate to form optic cups, and the retina develops from their inner walls. (C) The retinal layers in the adult. GCL, ganglion cell layer; IPL, inner plexiform (synaptic) layer; INL, inner nuclear (second-order cells) layer; OPL, outer plexiform layer; ONL, outer nuclear layer consisting of the cell bodies of the photoreceptors.

The degeneration of photoreceptors in a detached piece of retina begins within hours, and unless the retina is reattached within a week—often by laser surgery—photoreceptors are lost. Detached retinas are a serious clinical problem.

Another consequence of the retina forming as it does is that light must pass through the thickness of the retina to reach the photoreceptors. For most of the retina this is of little consequence for vision, as the retina is reasonably transparent, but for high-acuity vision it does make a significant difference. In the high-acuity (foveal) region of the retina, the inner layers of the retina are displaced aside, allowing light to impinge more directly on the photoreceptors as shown in figure 1.6, a drawing by Stephen Polyak, a giant in the field of primate retinal anatomy who worked mainly in the 1930s. The foveal region is quite small—only a half-millimeter across, but if the foveal area is damaged, such as occurs in age-related macular degeneration, visual acuity is severely compromised. One can no longer read, recognize faces, or drive a car.

The anterior eye structures—cornea and lens—form in an interesting way. The optic cup induces the lens to form from the overlying ectodermal cells, which otherwise would form skin cells, but as a result of the optic cup interacting with these cells, they proliferate, forming a structure called the lens placode, which breaks off from the overlying ectoderm and eventually

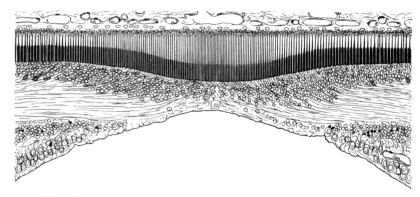

Figure 1.6
Drawing of the fovea in humans by Stephen Polyak. The inner layers of
the retina are displaced laterally so that light can impinge more directly
on the photoreceptors. Only cones are present in the central foveal region, and they are the thinnest and longest cones in the retina.

becomes the lens (figure 1.7). Subsequently, the lens induces the
corneal cells to form, also from ectodermal cells. The cornea and
lens are discussed further in chapter 2.

Organization of the Retina

The retina is one of the best-understood parts of the brain, and
all vertebrate retinas are organized with the same basic plan.
Three cellular (nuclear) layers containing the cell bodies of the
retinal cells are separated by two synaptic (plexiform) layers
within which the cells communicate with each other (figure
1.8). The photoreceptors (R), cones and rods, are the most distal
retinal elements; their specialized outer-segment regions, which
capture the light, are in close apposition to the overlying pigment epithelial cells.

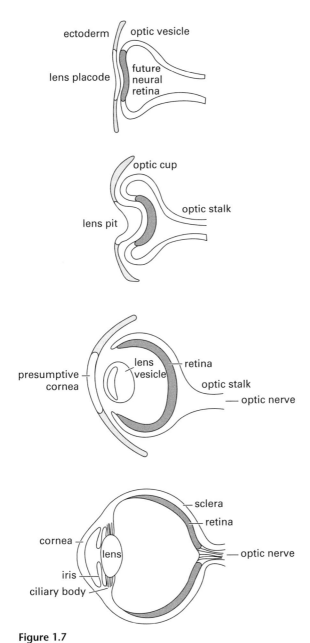

Figure 1.7
Formation of the lens and cornea. The optic cup contacts overlying ectodermal cells, resulting in the formation of the lens placode, which invaginates and forms the lens. The lens subsequently induces the cornea and other anterior eye structures to form.

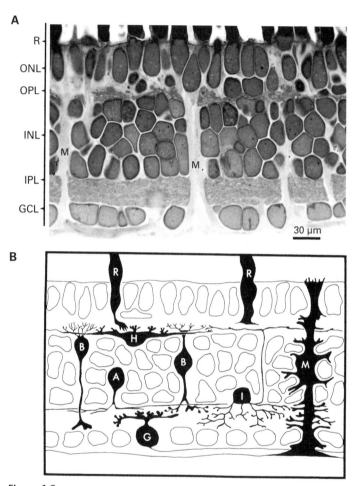

Figure 1.8
(A) A micrograph of a portion of a retina removed from the mudpuppy (an amphibian) eye. The mudpuppy has unusually large cells, which made it possible to record the electrical activity generated by the cells. As in all vertebrate retinas, three cellular layers (ONL, INL, GCL) and two synaptic layers (OPL, IPL) are seen. (B) Drawings of the retinal cells that reside in the cellular layers and contribute processes to the plexiform layers. R, cone and rod photoreceptor cells; H, horizontal cell; B, bipolar cells; A, amacrine cells; G, ganglion cell; M, Müller glial cell.

In the first synaptic layer, the outer plexiform layer (OPL), the rod and cone photoreceptors contact (synapse onto) two neuronal cell types: bipolar cells (B), which carry visual information from the outer to the inner retina, and horizontal cells (H), whose processes extend laterally in the OPL and mediate lateral interactions between photoreceptors and between photoreceptors and bipolar cells.

Bipolar cells provide the input to the inner plexiform layer (IPL) and synapse onto amacrine (A) and ganglion cells (G). The ganglion cells are the output neurons of the retina; their long axonal processes run along the inner surface of the retina, collect together to form the optic nerve, and carry the visual message to the rest of the brain (see figure 1.1). Amacrine cells, like horizontal cells, confine their processes for the most part to the IPL and mediate interactions between and among bipolar, amacrine, and ganglion cells. Some ganglion cells get much of their input directly from bipolar cells; others get input mainly from amacrine cells.

In addition to the neurons in the retina, there are also present nonneural cells called glial cells, which provide support to the retinal neurons. The principal glial cell in the retina is called the Müller cell, which extends across the entire thickness of the retina and whose processes envelop all of the neurons. Along the inner margin of the retina and where the optic nerve exits the eye is another type of glial cell—astrocytes—which are implicated in certain retinal diseases such as glaucoma.

The above description is simplified. There are multiple subtypes of each of the retinal neurons—perhaps as many as 18–20 amacrine cells in mammals, for example, and 12–15 types of ganglion cells depending on the animal. There is compelling evidence that 12–15 different messages concerning an image on the retina are constantly being transmitted to higher brain areas.

Visual Deficits and Blindness

Defects or disease in virtually any part of the visual system usually compromise vision. If light cannot gain entrance to the eye because of an opaque lens, or if an image does not sharply focus on the retina, vision is debilitated. Defects in the photoreceptors or retinal neurons can also result in visual loss, and if there is damage to higher visual centers beyond the eye, a variety of visual problems can occur.

If the defect or disease is sufficiently severe, the individual is described as having a blinding condition, but the term blindness is often used to describe a variety of visual deficits that do not render the person sightless. Vision is most commonly measured in term of visual acuity, the smallest letter or number on an eye chart that can be read. The chart was devised by Herman Snellen in 1868 and is still known as the Snellen Eye Chart. It consists of 11 lines of letters or numbers, with the largest figure at the top and each successive line smaller. This test is done routinely in virtually every eye examination and measures foveal vision, which mediates our high-acuity vision. Normal vision is called 20/20, which means that one can distinguish a letter or number approximately 8.5 mm high (about 1/3 inch) from 20 feet away, and normal eyes can readily do this. Twenty feet was chosen because the rays of light emanating from an object at that distance are parallel, as they are from an object at infinity. If a patient tested has less than 20/20 vision—say 20/40—this means that individual recognizes a figure 17 mm high at 20 feet that a normal individual recognizes at 40 feet. Some people, particularly when young, have better than normal vision and have acuities of 20/15 or even 20/10. Most commonly when normal acuity is not achieved, it is because of a refractive error—the image of the letter or number is not focused sharply on the retina. Refractive deficits can usually be corrected with glasses, and vision now becomes 20/20. On the other hand, a defect in the central retina, which houses the fovea, can compromise visual acuity, and such defects are impossible to correct. This is the case in age-related macular degeneration (AMD), which affects the fovea and the surrounding so-called macular and parafoveal regions of the retina. If, for example, corrected vision is less than 20/40 in both eyes, most states will not issue an unrestricted driver's license.

When corrected high-acuity vision is degraded to 20/200—the patient can see at 20 feet what a normal person sees at 200 feet—the patient is termed legally blind because he/she cannot read without aids, recognize faces, or do many routine visual tasks that require high acuity. On a Snellen eye chart, this means a legally blind individual can recognize only the large letter (usually an E) at the top of the chart, which is 10 times larger (~85 mm high) than the letters a person with 20/20 vision can distinguish. However, legally blind patients with deficits limited just to the fovea and closely surrounding regions, such as is usually the case in late-stage AMD, retain vision over the rest of the retina and can function visually to a considerable extent. They can see and avoid large objects, navigate visually, and perform other visual tasks that require only low-acuity vision. They are not totally blind.

Our highest acuity is in the fovea center, and acuity falls off very steeply away from the fovea. You can easily demonstrate this by taking two wood pencils that are banded just below the eraser (or similar objects). Focus your eyes on the left pencil and move the right pencil to the right. After a movement of only a few inches the banding (ferrule) is difficult to discriminate, and as you move further away, the banding completely disappears, although you can still see the pencil. To see the banding clearly you must move your eyes to bring the image of the pencil on the right onto the fovea, but then the banding on the left pencil is no longer seen.

The converse to the loss of foveal vision is seen in an inherited retinal degeneration, retinitis pigmentosa. This disease affects first the photoreceptors, especially the rod photoreceptors in the periphery of the retina. The fovea is spared at least initially, so the loss of vision is manifest by a restriction of the visual field. As peripheral vision is gradually lost, the visual field shrinks until only foveal vision is left. Visual acuity may still be normal (20/20) or almost normal, but the patient now has lost the ability to bring an object to focus on the fovea. In other words, the patient cannot locate visually the image he or she wishes to see. To do this requires peripheral vision. Termed gun-barrel vision, this condition is exceptionally debilitating. Indeed, an eye is also termed legally blind if its visual field is less than 20°, whereas in a normal eye the visual field is about 150°. But again these patients are not totally blind, although in most cases of retinitis pigmentosa these patients do eventually lose their central vision and become completely blind.

When rods are selectively compromised as in early retinitis pigmentosa, a patient is termed night-blind because to see in very dim light requires rods. Rods are more sensitive to light than are cones, and you can easily observe this yourself on a starry night by looking up until you see a dim star off to one side of where you are looking. If now you change your gaze to look directly at the dim star, it disappears because the foveal cones (and the central fovea contains only cones) are not sensitive enough to detect the dim star. If you look away, the dim star reappears, because its light is now falling on rods outside of the fovea.

Conversely, when a defect affects cones, one is termed color blind, and the most common color blindness defects occur as a result of the loss or alteration of one of the three cone photoreceptors humans have. But both central and peripheral vision in such people is quite normal—they have normal acuity and can bring objects to focus on the fovea—but they do not see colors as do normal individuals. Most color-blind individuals who have lost one cone photoreceptor type can still make color discriminations, and many color-blind people do not even recognize they have the defect.

To summarize, all of the above visual defects are termed "blindness" of one sort or another, but such patients still maintain considerable visual capabilities. On the other hand, if the transmission of visual information from the eye to the rest of the brain is lost, as happens in glaucoma—a disease that destroys the output retinal neurons, the retinal ganglion cells—then all vision is lost, and the individual is absolutely blind, not having any light perception or any visual capabilities at all. The bottom line is that most eye deficits that result in conditions described as blinding conditions result in compromised vision but not total blindness.

So far, we have focused on eye deficits, but deficits in higher visual structures can also cause blindness both partial and total. A large lesion in the LGN or in V1 of the cortex also can result in blindness. Such patients may have normal eyes, but without a functioning area V1 or LGN, they do not perceive images. These patients say they cannot "see" anything. However, not all of the retinal output goes to the LGN and on to the cortex. Some optic nerve axons project to other, noncortical higher visual centers that are involved in the regulation of pupil size, circadian rhythms, and other phenomena requiring visual input, and these visual functions are retained. It is even the case that these patients can respond to crude visual stimuli, but they cannot explain why they responded. As far as the patients are concerned, they are totally blind; they have no visual perception.

With lesions in higher cortical areas beyond area V1, the visual deficits the patients experience are often more limited and quite specific. For example, a small lesion in one cortical area results in the loss of ability to perceive any color, whereas a lesion in another area results in the loss of ability to see movement. A lesion along the "what" pathway described earlier results in patients unable to perceive objects or images, but such patients can "see" to avoid objects, although they do not perceive the objects. All of this is described in detail in chapter 6.

To conclude, various visual deficits reflect damage to or disease of specific cells or parts of the visual system, and careful characterization of a visual deficit can be instructive with regard to which cells or parts of the brain are involved and how those cells or parts of the brain contribute to vision. To understand the underlying disease mechanism or cellular alterations resulting in a visual deficit requires an understanding in some detail of how that cell normally works or how that part of the brain functions. Then one can begin to suggest underlying disease mechanisms and possible therapies to correct a deficit. This is the goal of the rest of the book.

2 Focusing Light—Cornea and Lens:

Refractive Errors, Dry Eye, and Cataracts

The eye is an extraordinarily intricate structure. Figure 2.1 is a schematic diagram showing the major components of the eye. Light rays enter the eye via the transparent cornea, which together with the lens focuses images on the photoreceptors in the back of the eye. The iris regulates the amount of light entering the back of the eye by dilating and constricting its central opening (the pupil, plate 1). The lens, located just behind the iris, is held in place by zonular fibers that insert into the ciliary body. Anterior to the lens is a watery fluid, the aqueous humor, which is responsible for maintaining eye pressure and shape. The aqueous humor is secreted by the ciliary body and circulates from the posterior to the anterior chamber through the pupil and then leaves the eye through tiny drainage channels located in the periphery of the anterior chamber.

Behind the lens is the vitreous humor, a gel that fills the vitreous cavity and helps to hold the retina in place. The retina and pigment epithelium line the back of the eye, and distal to them is the choroid, a highly vascular tissue that supplies the pigment epithelium and photoreceptors and other outer retinal cells with oxygen and nutrients. The outer coat of the eye is the tough white sclera. The optic nerve, consisting of the axons of the ganglion cells, also contains the central retinal artery and vein. The

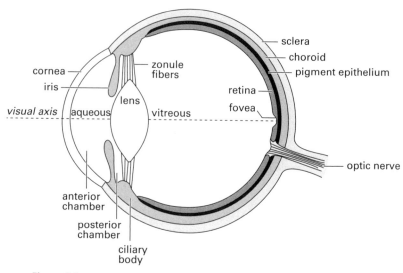

cornea
iris
visual axis
aqueous
lens
zonule fibers
vitreous
sclera
choroid
pigment epithelium
retina
fovea
optic nerve
anterior chamber
posterior chamber
ciliary body

Figure 2.1

Schematic diagram of the major components of the human eye. See text.

central retinal artery, on entering the retina, divides into smaller and smaller blood vessels that provide oxygen and metabolic support primarily for the inner retina (plate 5). The indented fovea (figure 1.5) is positioned on the optical axis of the eye and, as discussed earlier, mediates our high-acuity vision.

The cornea and lens are responsible for the focus of the eye; that is, they bring the rays of light emanating from objects at different distances to a sharp point on the photoreceptor cells. The cornea has greater focusing power than does the lens. Indeed, about two thirds of the focusing is accomplished by the cornea, but it is a fixed focus. The focusing power of the lens can be changed by muscles in the ciliary body that can alter the shape of the lens by changing tension on the zonular fibers. This

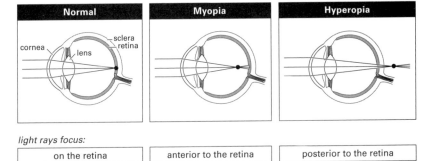

Figure 2.2
The focusing of light by the cornea and lens in the normal eye (left) and
in the common refractive errors (middle, right).

process, known as accommodation, enables the eye to focus
clearly on objects that are at different distances.

The major cause of impaired vision throughout the world is
refractive error, the situation in which the focus of the eye is not
on the retinal photoreceptors (figure 2.2). The refractive errors are
nearsightedness (myopia), farsightedness (hyperopia), astigmatism, and presbyopia. Myopia is the condition in which the rays
of light come to focus in front of the photoreceptors. It is caused
by excessive curvature of the cornea or an excessively long axial
eye length. In hyperopia the rays of light are destined to focus
beyond the photoreceptors—the result of a small, short eye or a
flatter cornea. Hyperopes can see more clearly at distance than
near. Severely myopic people who have very blurred distance
vision often believe they have "weak eyes" because they cannot see distant objects well. Actually most myopes have strong
healthy eyes—glasses can usually correct their vision to 20/20.
As myopia progresses in childhood and distance vision becomes
more blurred, the cause is not weakness but increasing growth
of the eye. The take home lesson from this is that only healthy

eyes can see clearly with glasses; a patient with a structural eye disease usually cannot see clearly, even with glasses.

Astigmatism is the situation in which the cornea is not spherical, and rays of light entering the eye on one plane focus at a different point than rays entering on a different plane (the cornea is shaped more like the end of an American football than the side of a basketball). Presbyopia (characteristic of aging) is not due to a weakness of the focusing muscles or an alteration in the shape or structure of the eye, unlike other refractive errors, but to a loss of elasticity of the lens with age so that the focusing muscles cannot change the shape of the lens effectively. This situation usually manifests in the mid-40s; focus for near objects is compromised, and it becomes increasingly difficult to see small objects such as newsprint, so reading glasses are required. Far-sightedness and near-sightedness are frequently confused, but the terms are as described. Far-sighted individuals see better at a distance, whereas those near-sighted see better close.

The management of refractive errors is largely accomplished with glasses, which can be fashioned to bring the focus to the appropriate point on the retina. Astigmatism is corrected by lenses known as cylinders, which focus rays of light on two different planes. Presbyopia requires magnifying lenses of gradually increasing strength added to any basic refractive error. Aging myopes can usually see near objects clearly without glasses but still require glasses for distance. Those in the presbyopic age group (45 years and beyond) who require distance glasses need a different prescription for near vision—a problem that is solved by two separate pairs of glasses or by multifocal lenses, which may be bifocals, trifocals, or progressive lenses. Bifocals, as the name implies, have two foci, one for distance and one for near but have no focus for intermediate distances. Trifocals have three foci—near, distance, and intermediate focus, whereas progressive lenses increase in strength gradually from top to bottom and have continuous focusing capacity serving all distances.

Myopia is the most common refractive error. The cause of myopia is a classic example of the long-standing nature-versus-nurture debate. We know that genetic makeup—nature—is an important factor. Indeed, most myopes have a family history of myopia. Identical twins are usually both myopic or not myopic as compared to fraternal twins, where only one sibling may have myopia. The incidence of myopia varies throughout the world but is consistently highest in East and South Asian populations. As expected, hyperopia is more common where the incidence of myopia is lowest. In childhood and through adolescence, myopia gradually increases in incidence and degree as hyperopia decreases, a fact correlated with axial growth of the eye.

The childhood increase in the degree of myopia has long been thought related to increasing axial length associated with eye growth, but there is also evidence that environmental factors such as prolonged near focusing influence eye growth and myopia—that is, nurture. For example, 70% to 80% of students in urban Taiwanese schools are myopic compared to just 20% to 30% of comparable children who live in rural areas. Further, it has been shown that raising monkeys in a confined space—a small cage, for example—induces myopia.

In the 1970s Elio Raviola and Torsten Wiesel observed that raising monkeys with one eye closed, thereby preventing sharp images from falling on the photoreceptors, also induced a severe myopia in the deprived eye (deprivation myopia). In a variety of animals it has been shown that lenses that alter where an image is focused on the back of the eye, either in front of or beyond the retina, can change the shape of the eye and cause refractive errors. These experiments suggest that an intrinsic mechanism within the eye, deriving from the retina, can alter the growth of the eye. It is as if the eye is seeking to optimize focus and thus grows or shortens in length, becoming myopic or hyperopic. How might this come about? Raviola and colleagues went on to show that in an animal made myopic, there is a considerable increase in a small protein, a neuropeptide called vasoactive intestinal peptide (VIP), within one type of amacrine cell. The suggestion has been made that, under conditions that induce myopia in animals, the expression of VIP is increased in this particular amacrine cell. Subsequently it is released from the cell and initiates retinal and eye growth. Another substance found in retinas that may be involved in eye growth is dopamine. That eye growth appears to be controlled by an intrinsic retinal mechanism has also been demonstrated by the fact that deprivation myopia still occurs in animals after the optic nerve has been severed, indicating that central visual mechanisms are not involved. However, much remains to be learned about the control of eye axial length and the development of myopia.

The shape of the cornea (corneal curvature) can be changed by various refractive surgical procedures (such as LASIK) to bring rays of light to a clear focus on the retina, eliminating the refractive error and the need for glasses. Corneal shaping procedures are particularly effective in myopes, but when such patients reach the age of presbyopia, they will need to wear glasses for near vision. To eliminate glasses completely requires attention to the lens of the eye. A lens can be replaced by an implant lens that can be progressively accommodating; that is, when looking straight ahead, the lens is focused for far vision, when looking down, the lens brings near objects into focus. This procedure is commonly incorporated in cataract surgery but in rare cases is performed on individuals who have a severe refractive error but whose lenses are clear.

Cornea

The cornea is a thin, transparent tissue continuous with the sclera and contains no blood vessels. Indeed, under normal conditions, blood vessels are actively prevented from entering the cornea, likely by proteins released from corneal cells that inhibit blood vessel formation. Only when the cornea is damaged or diseased do blood vessels grow into the cornea, and when this happens, a process called neovascularization, vision can be severely compromised. Without a blood supply, the cornea receives its nourishment from tears on its outer surface and the aqueous humor on its inner surface.

The cornea consists of three major layers (figure 2.3): an outside layer of epithelial cells that provide a barrier preventing bacteria and other foreign substances from access to the internal corneal layers; a thicker, middle (stroma) layer containing few

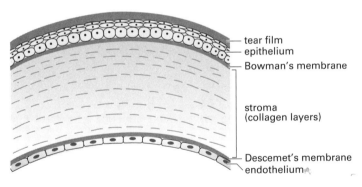

Figure 2.3
Schematic drawing of the cornea showing its five layers and the tear film on its surface. The stroma consists mainly of collagen fibers arranged in layers orthogonal to one another (see figure 2.4).

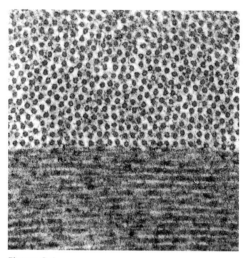

Figure 2.4
Transmission electron micrograph of the corneal stroma in a human
cornea. The fibrils are shown in cross section in the upper half of the
micrograph. Fibrils arranged longitudinally are seen at the bottom.

cells but numerous long thin protein fibers called collagen, that
are precisely and tightly arranged in parallel bundles (figure 2.4);
and an inner endothelial cellular layer that is mainly responsible
for maintaining the transparency of the cornea.

To maintain and nourish the outer epithelial cells, a tear film
continuously bathes the surface of the cornea, the nature of
which is described in more detail later. The corneal epithelium is
also richly innervated with sensory nerve endings. Thus, injury
to the cornea is especially painful. But it also means that even a
light touch to the cornea elicits the corneal reflex—a rapid closing of the lids that provides protection to this critical tissue.

Two other corneal structures exist—Bowman's membrane,
which sits between the epithelial cells and stroma and plays a

role in securing the epithelial cells to the underlying stroma, and Descemet's membrane, a thinner structure that contains collagen fibers along with other molecules. Bowman's membrane is formed by the epithelial cells, whereas Descemet's membrane derives from the endothelial cells.

The main function of the endothelial cell layer is to maintain the transparency of the cornea. If the endothelial cells are compromised by disease or damage, the cornea swells and becomes cloudy. To maintain a transparent cornea, the endothelial cells need to keep the cornea in a partially dehydrated state.. When the endothelial cells are damaged, water enters the stroma, separating the collagen fibers and causing swelling and clouding of the cornea.

The endothelial cells maintain the cornea in a dehydrated state by two mechanisms. First, the cells form a continuous layer under the entire cornea, and the endothelial cells tightly adhere to one another laterally via specialized junctions called tight junctions (figure 2.5). These junctions substantially decrease the flow of fluid between the cells and thus provide a barrier between the stroma and the aqueous humor below. Second, and more important, along the lateral sides of the cells are numerous protein pumps that continuously extrude ions, especially sodium (Na^+) and bicarbonate ions (HCO_3^-), from the stroma into the aqueous humor below. As the number of ions decreases in the stroma, water follows, keeping the stroma dehydrated with only about three-quarters the amount of water it would have without the action of the pumps. We experience the same phenomenon when we eat salty nuts or potato chips. The increased salt in the blood and extracellular fluid draws water from our cells, and we become thirsty. Drinking water dilutes the salt, and water flows back into the cells, eliminating our thirst.

Each endothelial cell has about 3 million of these protein pumps, resulting in a pump density of 4.4 trillion per square

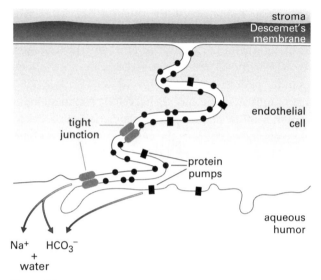

Figure 2.5
Diagram of the junctions between corneal epithelium cells and the location of the protein pumps that extrude Na^+ and HCO_3^- from the cells.

millimeter along the lateral membrane of an endothelial cell, an incredible number. The pumps require energy, which they obtain from the energy-rich molecule adenosine triphosphate (ATP). The endothelial cells are some of the most metabolically active cells in the eye and body.

What causes endothelial cell failure? As one ages, the number of endothelial cells decreases by as much as 50% over the course of one's life, but the safety factor is considerable. Indeed, one needs to lose about 90% of the endothelial cells before corneal edema occurs, and this would take something like 250 years at normal rates of endothelial cell loss. However, there are specific diseases that affect the endothelium, the most common of which is known as Fuchs' dystrophy, which occurs later in life and whose cause is unknown.

Depending on their extent and density, corneal opacities can have a drastic effect on vision. The cornea is also subject to a variety of other pathologies, particularly trauma but also congenital abnormalities, metabolic disorders, the effects of aging (in addition to the loss of endothelial cells), and infection by a variety of agents including bacteria, viruses (particularly herpes simplex), fungi, and parasites. A common sequela of corneal disease is scar formation and cloudiness, which can severely impair visual capability.

To restore clarity, corneal transplants are surgically performed, and this procedure has become routine with a rejection rate today of only 10% after 2 years. First carried out in 1905 by Eduard Zirm, an Austrian, the cornea was the first solid organ to be transplanted successfully from a donor, and today some 34,000 corneal transplants are performed annually in the United States. The traditional technique was to remove a full-thickness button of the cornea encompassing the pathology and replace it with a comparable full-thickness graft from the cornea of a donor. These penetrating keratoplasties (as they are called) are reasonably successful, but the postoperative visual acuity is frequently less than ideal, and a variety of complications do occur including infection and hemorrhage as well as the initiation of glaucoma, cataract, excessive astigmatism, and even graft failure. Because corneas do not have blood vessels, rejection reactions are generally mild—which is why corneas were the first successful organ transplant—and today most rejection reactions are well controlled by topically applied drugs.

In recent years remarkable advances in corneal surgery have occurred. Instead of the traditional full-thickness button of donor cornea, the newer procedures transplant a replacement for only the diseased part of the cornea. These operations, known as selective corneal transplantations, retain the uninvolved, healthy parts of the cornea. For example, a cornea with a dense scar in the anterior one-third is treated with an anterior donor corneal inlay, whereas a scar in the posterior one-third is treated with a posterior inlay. These procedures may not involve entry into the eye, which significantly diminishes the number and severity of potential complications. A cloudy, edematous cornea from an abnormality of the endothelium is treated by careful removal of Descemet's membrane and diseased endothelium and replacement with a donor Descemet's membrane and endothelium. These procedures have also significantly improved the level of vision over that obtained by the traditional penetrating keratoplasty. Selective corneal transplantation has revolutionized corneal surgery and continues to improve almost day by day.

Another recent development in the treatment of corneal abnormalities involves the relatively common corneal condition known as keratoconus. The center of the cornea is excessively thin in keratoconus and gradually protrudes forward in a cone shape. Usually seen in young people, it progressively and severely impairs vision. Traditionally it has been managed in its early stages by glasses and contact lenses but, when severe, by penetrating keratoplasty surgery. A new procedure, called corneal collagen cross-linking, involves an interaction among ultraviolet light, vitamin B_2, and collagen protein. This procedure, which strengthens the cornea and prevents progression of the corneal distortion, is widely used in Europe and is currently under review by the FDA regulatory agency in the United States.

Some corneas are in such poor condition overall that standard transplant surgery has little chance of success. For these patients an artificial, plastic cornea has been developed with good results in patients who have no other options for restoration of their vision.

Dry Eye

A common and distressing corneal disorder is dry eye, in which the cornea is not bathed properly by tears, and there occurs ocular surface discomfort and occasionally decreased vision. As many as 10–15% of the adult population are affected. It is more common in women than men, and symptoms range from minor corneal irritation—itching and grittiness—to a burning sensation and photophobia (sensitivity to light). If severe enough, infection and inflammation can occur, and even corneal ulceration and perforation. Thus, dry eye, if not dealt with, can have serious consequences, although this is rare.

Dry eye is caused by disruption or deficiency of the tear film that bathes the corneal surface. Normally we blink every 5–10 seconds, renewing the tear film, and if blinking is delayed by more than 10 seconds, breakup of the tear film begins to occur. People with dry eye, on the other hand, may have their tear film break up in just 1–2 seconds.

The tear film is more than an aqueous layer. Indeed, it consists of three separate layers: an inner mucous layer that acts as a wetting agent, a thicker middle aqueous layer, and on the outer surface of the tear film, a thin lipid (fatty) layer that retards evaporation (figure 2.6). These three layers are secreted by different glands found underneath or in the eyelids (figure 2.7). The aqueous layer, which consists of ions, glucose, and various proteins as well as water, is mainly produced by the prominent lacrimal glands, whereas the mucous layer is made by adjacent conjunctival goblet cells. In the mucous layer are large proteins called mucins that form a gel. The lipid layer is secreted by Meibomian gland cells found on the edges of the eyelids. Secretions from the gland cells are mainly under the control of sensory nerves found in the cornea that activate nerves innervating the glands.

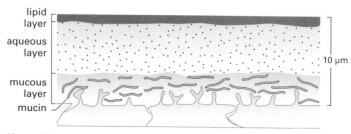

Figure 2.6
The tear film consists of three layers: an outer lipid layer that retards evaporation, a middle aqueous layer, and an inner mucous layer. The underlying epithelial cells on the corneal surface are flattened and extend fine processes into the mucous layer.

Any disruption of the glands, secreting cells, innervating nerves, or ducts that secrete the fluids on the corneal surface can cause dry eye, including age, infection, and trauma. Treatments include lubrication and hydration of the corneal surface with artificial tears as well as drugs that enhance secretion from the various glands, especially the lacrimal gland. Another treatment is to block the outflow channels for tears, to preserve what tears are produced. Unfortunately, the various agents presently used are costly, not enormously effective, and must be applied frequently.

Lens

The lens, like the cornea, is also highly transparent, as it must be if it is to transmit light efficiently. Unlike the cornea, which is fixed in structure, the lens is somewhat malleable; that is, its shape can change to allow for accommodation. Thus images near and far can be focused on the photoreceptors by shape changes of the lens (figure 2.8). When the ciliary muscle contracts,

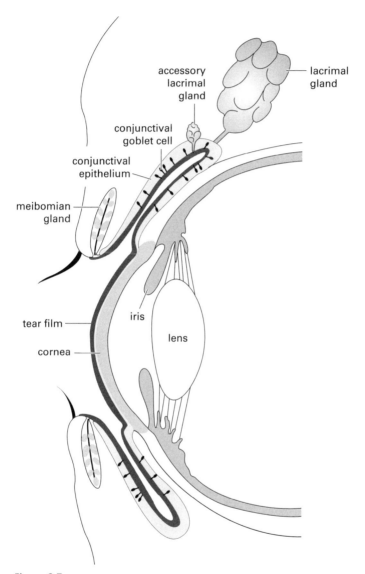

Figure 2.7
Schema of the glands in the eye lids that secrete the tear film. See text.

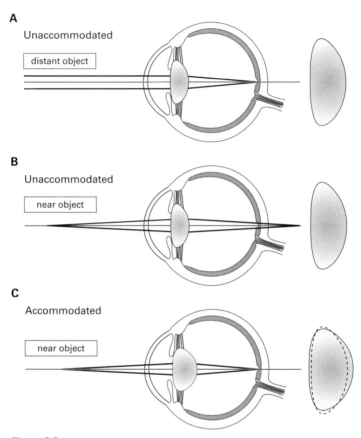

Figure 2.8
Accommodation (A). In the unaccommodated state, the lens and cornea focus far objects on the retina, whereas (B) the focus for near objects is behind the retina. With accommodation, the lens becomes rounder and thicker, moving the focus forward onto the retina (C).

tension on fine fibers (called zonular fibers) that attach the lens to the ciliary body decreases (see figure 2.1). When changing focus from a far to a near object, the tension decrease on the zonular fibers allows the lens to become rounder and thicker—to become a stronger lens in other words.

Whereas corneal transparency is due mainly to an acellular stroma, the transparency of the lens is due primarily to highly modified lens cells. That is, during differentiation lens cells lose their nuclei and most of their other organelles while they increase their protein concentration substantially, from 15% to 70%. The major proteins in differentiated lens cells, which are called crystallins, do not scatter light as do most proteins. They are relatively homogeneous and stable proteins and pack together tightly. They are water soluble, and in lens cells they form a viscous matrix. The lens cells, when differentiated, are called lens fiber cells, and they too are closely packed in precise ways.

Interestingly, Grahame Wistow and Joram Piatagorsky in the 1980s showed that many of the lens crystallins derive from and are closely related to proteins found elsewhere in the body including a number of enzymes, proteins that mediate specific biochemical reactions within cells. In some cases the lens crystallins even retain enzymatic activity, as can be shown by incubating certain crystallins with appropriate substrates and observing the biochemical reaction. The lens thus conscripts proteins from elsewhere that have the necessary properties to allow for transparency. This phenomenon is known as protein or gene sharing.

The lens derives from the ectoderm overlying the optic vesicles, and it is the optic vesicles that induce these ectodermal cells to become lens cells. This induction is mediated by specific proteins secreted by the optic vesicles. The induced cells first elongate slightly, forming a lens placode, which then invaginates, forming a lens vesicle. The most posterior cells in the vesicle are the first to differentiate into lens fiber cells; they elongate substantially and fill the vesicle (figure 2.9). The more anterior epithelial cells extend down to the equator of the lens, elongate, and form secondary lens fiber cells that surround the primary lens fiber cells. The epithelium proliferates as the lens increases in size to maintain an epithelial layer that extends approximately to the equator of the lens. Indeed, lens epithelial cells continue to proliferate in a narrow band around the equator, and the lens continues to grow throughout life, although the growth becomes progressively slower with age.

Once a lens fiber cell has formed, it persists and is not replaced. Because these cells have lost most of their organelles, they must be supported, and if the cells are not properly maintained, pathology develops. Thus, the initial lens fiber cells may be the oldest cells extant in the human body. Lens fiber cells formed before sexual maturity are grouped in the center of the lens and are identified as the lens nucleus, whereas lens fiber cells formed later are called cortical lens cells.

Because, as in the cornea, there are no blood vessels in the lens, how are the cells maintained? First, since they have lost virtually all of their organelles, their metabolism is very low—no

Figure 2.9
Diagrammatic representations of key structural events in lens development. From the top: the thickened lens placode overlying the optic vesicle; the invagination of the lens placode into the developing optic cup; the hollow lens vesicle and the onset of primary lens fiber cell elongation of posterior lens cells; progressive obliteration of the lens vesicle lumen by elongating primary lens fiber cells; the embryonic nucleus composed of the primary lens fiber cell mass and the overlying anterior lens epithelium; the human lens at birth, with its nucleus composed solely of primary lens fiber cells surrounded by secondary lens fiber cells.

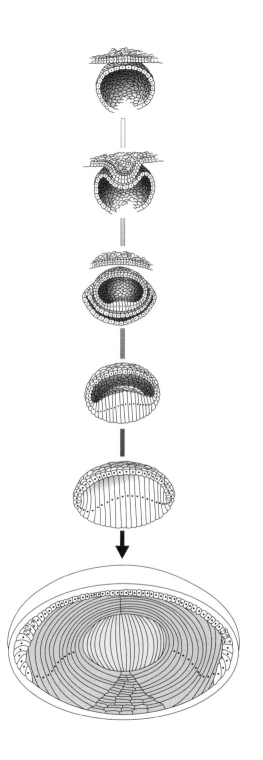

protein synthesis or gene expression takes place. They derive energy from the breakdown of glucose to lactic acid, a process known as glycolysis, which does not require oxygen. It is presumed that glucose and other required metabolites reach the cells via simple diffusion from the aqueous humor. Without protein turnover, it does mean that lens proteins must be extraordinarily stable, lasting up to 100 years or more.

The lens fiber cells are closely packed together in a precise lattice structure. There is little space between the cells, and there are no blood or lymphatic vessels in the lens. If the lattice is disturbed, lens transparency declines, so lens transparency depends on both the high intracellular concentration of crystalline proteins and the arrangement and packing of the lens fiber cells themselves. The other structure of note is the lens capsule, a relatively tough fibrous structure surrounding the lens and made up of proteins secreted by the epithelial cells. The zonular fibers responsible for accommodation insert into the capsule but originate in the ciliary muscle found in the ciliary body.

Cataracts

Cataract, defined as opacity of the lens affecting vision, is the most frequent cause of blindness in the world. As many as 30 million people worldwide may have their vision severely compromised by cataracts. The major risk factor for cataracts is age with an exponential increase in cataract incidence after the age of 50. Other risk factors include sex—women are more cataract prone than men—as well as smoking and alcohol consumption. Poor nutrition is a risk factor, as is diabetes. Ionizing radiation exposure, including x-rays, can result in cataract formation, and there is evidence that ultraviolet (UV) light, especially short-wavelength UV, is a precipitating factor. The lens strongly absorbs UV light—indeed, the human lens sharply cuts off UV light from entering the eye, presumably to shield the retina from the potentially damaging effects of those wavelengths of light. Interestingly, some animals, especially fish, possess UV-sensitive photoreceptors, and their lenses do transmit UV light. Young zebrafish actively avoid UV light, whereas they are attracted by visible light (that is, light visible to us). If young zebrafish are raised in visible light that also contains some UV light, their mortality rate increases significantly, presumably because of the damaging effect of UV light.

The most common cataracts occur in the nucleus of the lens, which is composed of the oldest lens cells (plate 2). There is substantial evidence that nuclear cataracts are the result of oxidative damage to lens proteins, causing them to aggregate and scatter light. The lens fiber cells do contain antioxidation substances, but these seem to decrease with age. If levels of oxygen are increased near the lens as a result, for example, of surgery in which vitreous humor is removed, there is a clear increase in cataract formation. On the other hand, in diabetic retinopathy where oxygen levels in the eye are often reduced, there is a decreased incidence of cataract.

There are also cataracts in the younger, cortical regions of the lens. These cataracts are usually characterized by disruption of the lens fiber cells, something that is not seen in nuclear cataracts. The supposition is that the cell disruption is caused by physical damage to the lens cells, perhaps, by the hardening of the lens nuclear region. Cortical cataracts, being on the margin of the lens and not in the lens center, cause less vision loss than nuclear cataracts. Interestingly, nuclear cataracts tend to progress very slowly and can alter the focus of the lens, inducing a myopia. Thus, older individuals who previously needed reading glasses find they can now read comfortably without glasses—sometimes referred to as second sight.

Essentially every normal lens develops some lens opacities and darkening by age 50 or 60. These are usually cortical opacities in the lens periphery, which have little effect on vision and often do not progress over many years. When these opacities are seen by eye specialists, the patient is often told that he or she has the beginnings of cataracts, which causes worry even though these opacities frequently never significantly affect vision and do not require surgery. One type of cataract (posterior subcapsular) does tend to develop quickly, and the patient can go from good vision to very poor vision in only a few months, but this is an uncommon form of cataract. Many early lens opacities show no significant change over many years.

Congenital cataracts also occur. Infectious agents in pregnancy, such as rubella, the German measles virus, is one such agent. One feature of these cataracts is that during eye development the lens fiber cells do not degrade their cellular constituents as they differentiate; thus, the cells scatter light significantly, contributing to the lens opacification. There are also known mutations in lens-specific proteins that cause cataracts.

The major symptom of cataract formation is a gradual and painless blurring of vision, and there is currently no known way to prevent cataracts. The only treatment is surgical removal of the cloudy lens. Approximately 3 million cataract operations are performed in the United States alone each year, the most common surgical procedure in those over 65. The results of this surgery are generally excellent in restoring clear vision, as long as the eye is otherwise healthy and functional.

Cataract surgery is one of the oldest known surgical operations. For centuries the procedure consisted of knocking the lens out of its zonular attachments (a procedure called couching) and letting it sink to the bottom of the vitreous cavity. This provided bright but unfocused images onto the photoreceptors.

A significant advance, pioneered by Jacques Daviel in 1748, was to make an incision in the eye and to physically remove the cataractous lens. Following this procedure, the patient required very strong glasses (so called coke-bottle glasses) to focus. These cataract glasses also did not provide normal peripheral vision and were difficult to get used to, but they were far better than being essentially blind.

Cataract surgery took a huge step forward when Harold Ridley in England introduced a plastic lens as replacement for a cataractous lens in the 1940s. Ridley was an ophthalmologist who treated a fighter pilot during the Second World War who had a plastic fragment from his plane's windshield embedded in his eye. Ridley was impressed as to how well the eye tolerated the fragment and came up with the idea of a plastic implant to replace a cataractous lens. Initially ridiculed by the ophthalmic community, his idea took hold as the implants improved and eventually he was able to restore excellent vision in many patients. Today, implants are inserted routinely and provide appropriate focusing so patients can see normally near and far without glasses. If a patient has cataracts in both eyes, he or she can choose to have one lens for far vision, the other for near vision, and many patients find this useful. Interestingly, most patients quickly adjust to having a lens for near vision in one eye and a lens for far vision in the other eye, and say they are unaware of it. This means they seamlessly use one eye for near vision and the other eye for far vision, and the fact that the two eyes focus differently is not a problem.

A major question facing the ophthalmologist and patient is when a cataract should be removed. For many years it was standard procedure to wait until the cataract was "ripe." A "ripe" cataract is one in which the entire lens is opaque (and vision is essentially nil). The reason for that requirement was that with the surgical methods of the time, the results were better and complications fewer with complete cataracts. With current surgical techniques, ripeness is no longer a factor, and cataracts can be removed safely at any stage in their development. Currently, cataract surgery is recommended when a patient's vision with best glasses correction affects the person's quality of life; that is, impairs the patient's ability to drive, read, or carry out other usual visual activities.

3   Capturing Light—The Photoreceptors:
Retinitis Pigmentosa and Age-Related Macular Degeneration

As noted earlier, the light-sensitive cells in the eye, the photoreceptors, line the back of the eye and sit adjacent to the pigment
epithelium (figures 1.1, 1.5, and 2.1). Virtually all vertebrate
retinas have two types of photoreceptors, rods and cones. Rods
mediate dim-light (night) vision, whereas cones function in
bright light and mediate sharp central vision and color vision.
A retina without rods is night blind, but a retina without cones
is functionally blind; that is, we use our cones for virtually all
visual tasks, especially those requiring high acuity. Usually there
is just one type of rod in a retina, but there are several types
of cones. Humans, for example, have three types of cones that
respond maximally to red/yellow (long-wavelength), green
(medium-wavelength), or blue (short-wavelength) light (plate
4) and that underlie color vision, as first proposed by Thomas
Young in 1802. Other species have additional cone photoreceptors; many fish, for example, have ultraviolet-sensitive cones
as well as red-, green-, and blue-sensitive cones. What we "see"
depends on the wavelength sensitivity of our photoreceptors,
and so fish can see wavelengths of light that we cannot.

Both rods and cones are elongated cells with a specialized outer-segment region where light is captured, an inner-
segment region that contains various cellular organelles that

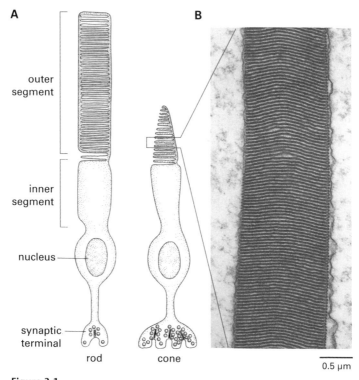

Figure 3.1

(A) In both rod and cone photoreceptors, the light-sensitive visual pigments are found in membranous disks in the outer segments of the cells. (B) A portion of a cone outer segment of a lizard. The lizard cone outer segments are long, and over a short portion of their length the cone shape is not obvious.

make energy-rich molecules such as ATP, or synthesize proteins and numerous other cellular constituents (figure 3.1). The photoreceptor nuclei containing the cell's genetic material (DNA) typically sit below the inner segments in the outer nuclear layer, and below that are the synaptic terminals where the visual signals from the photoreceptors are passed on to second-order bipolar and horizontal cells. In many species the cone outer segments are tapered and shorter than the rod outer segments (figure 3.1A). The names, rod and cone photoreceptors, derive from the shape of the outer segments.

The outer segments capture light by means of special light-sensitive molecules. These molecules are embedded in thin disk-like structures made up of membranes that derive from the outer membrane surrounding the outer segment. There may be as many as 1000 such membranous disks per outer segment (figure 3.1B), and each disk may contain as many as 100,000 light-sensitive molecules.

An interesting feature of both rod and cone photoreceptors is that they are continually replacing their outer segments, first demonstrated by Richard Young in the 1960s. New outer-segment material is synthesized in the inner segment and added to the base of the outer segment to form new disks. To keep the outer-segment length constant, distal outer-segment disks are shed daily and taken up by the adjacent pigment epithelial cells and digested there. This essential function of the pigment epithelium helps to maintain the photoreceptor cells. If this process is disrupted as a result of a gene mutation, for example, shed outer segments accumulate between the photoreceptors and pigment epithelial cells causing the death of the photoreceptors, probably by disrupting the flow of O_2 and nutrients to the photoreceptors from the vascular bed (choroid) beneath the pigment epithelium. In mammals, rod outer segments are completely replaced in about a week.

Visual Pigments

The light-sensitive molecules found in the outer segments of the photoreceptors are called visual pigments. They are called pigments because they absorb certain wavelengths of light but not all and are, therefore, colored. For example, the visual pigment of rods is called rhodopsin, and it absorbs green light best. It captures red and blue light less well, and it appears purple. As first shown in the late nineteenth century by Franz Boll, if a retina is removed from a dark-adapted animal such as a frog that has many large rods, the retina displays a reddish-purple color, which is why the original name for rhodopsin was visual purple (plate 3). The visual pigments of cones are chemically similar to the rod pigment rhodopsin, but they are maximally sensitive to other wavelengths of light.

There are two further points. First, the retina needs to be dark-adapted for the rhodopsin to be seen because light breaks down the visual pigment (a process called bleaching), and its color is lost. Second, the rhodopsin is packed into the outer segment at a very high density. A single rod outer segment contains between 10^8 and 10^9 molecules of rhodopsin. This means that 80–90% of the protein in an outer segment is visual pigment protein. The high density of visual pigment in an outer segment ensures that photons entering an outer segment are likely to be captured, and, further, the high density allows the color of the visual pigment to be seen. Most animals have many more rods than cones. Humans, for example, have about 120 million rods but only about 6 million cones.

All visual pigments, whether rod or cone, consist of a large protein (called opsin) to which is bound a modified form of vitamin A, as originally proposed by George Wald in the 1930s. The structure of vitamin A is shown in figure 3.2 along with its modified form that is bound in the protein to make a visual pigment. The modified form of vitamin A is called retinal, and the only change is that it has two fewer hydrogen (H) atoms at the far end of the molecule—its terminal group is an aldehyde (–CH=O) rather than an alcohol (–CH$_2$-OH). Figure 3.2 shows something else that is essential for photoreception. In the lower panel of the

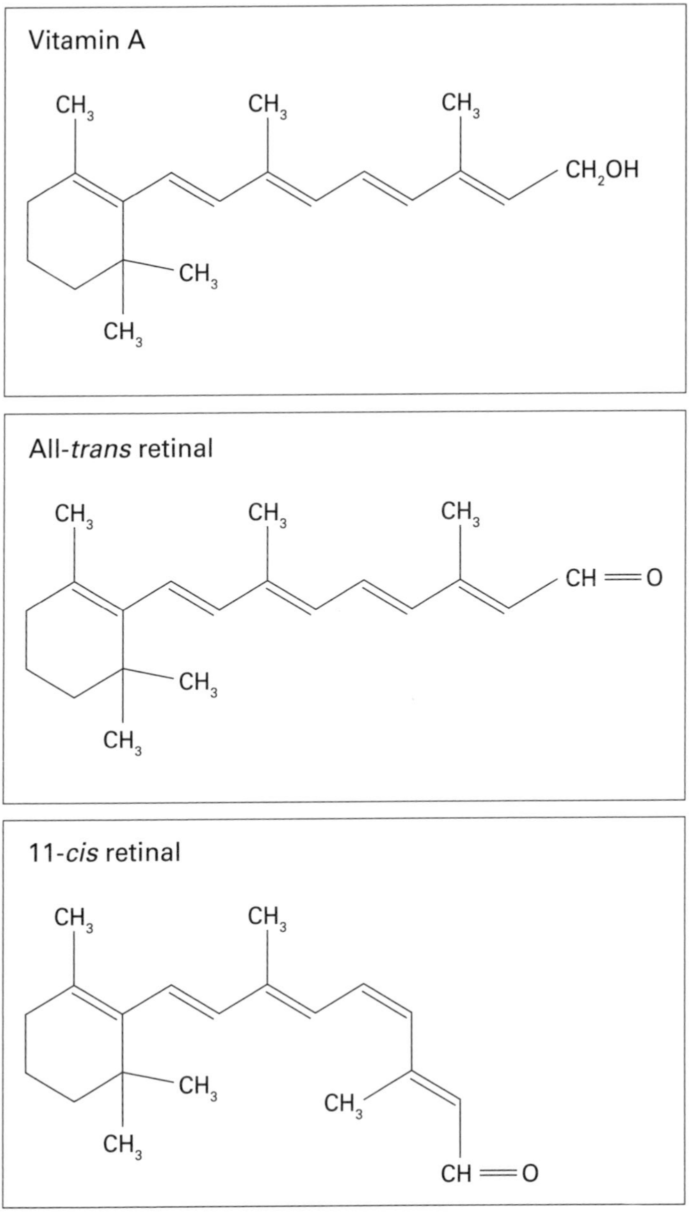

Figure 3.2
Structures of vitamin A, all-*trans*-retinal, and 11-*cis*-retinal. At every bend of the chain in the various molecules is a carbon atom with associated hydrogen atoms or a methyl group (CH₃).

figure is a molecule that has the identical atoms as the molecule shown in the middle panel, but it is bent. It is called an isomer, and to make a visual pigment molecule requires a bend in the appropriate position in retinal. Bent molecules are called *cis*-isomers, and one particular *cis*-isomer, 11-*cis*-retinal, is required to make all visual pigments.

This is important because it is the retinal that absorbs photons, making the molecule light sensitive. Further, when light is absorbed by retinal it isomerizes the retinal—from the 11-*cis* isomer to the all-*trans* form, again work from the Wald laboratory in the 1950s—and this is the entire role that light plays in the visual process!

Two things of consequence result from the isomerization of retinal. First, it allows the protein to change its shape, and when the opsin assumes a specific shape it becomes active, initiating a series of biochemical changes in the outer segment that lead to activation of the photoreceptor cell. Second, the all-*trans*-retinal is released from the protein and is converted back to vitamin A.

To remake or regenerate the visual pigment molecule, the vitamin A (in the all-*trans* configuration; top panel in figure 3.2) must be isomerized to 11-*cis*-vitamin A and converted back to 11-*cis*-retinal. This isomerization does not take place in the photoreceptor outer segments but in the adjacent pigment epithelium. Isomerization is a second critical role that the pigment epithelium plays in supporting photoreceptors, particularly the rods (cones differ in that much of their 11-*cis*-retinal is made in the Müller [glial] cells). Figure 3.3 shows the visual cycle. After exposure to light, isomerized (all-*trans*) retinal separates from the protein opsin, is converted to vitamin A, and is transported into the pigment epithelium, where it is isomerized to the 11-*cis* form. The 11-*cis*-vitamin A is converted to 11-*cis*-retinal and transported back into the outer segments, where it spontaneously combines with opsin to re-form rhodopsin.

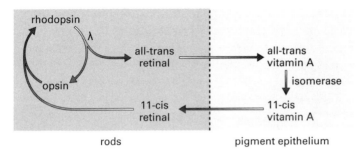

Figure 3.3

The essence of the visual cycle. Rhodopsin bleaches in the light to opsin and all-*trans*-retinal. All-*trans*-retinal is converted to all-*trans*-vitamin A, which is then transported to the pigment epithelium and isomerized via an enzyme (isomerase) to 11-*cis*-vitamin A. The 11-*cis*-vitamin A is transported back to the rods and converted to 11-*cis*-retinal, which spontaneously combines with opsin to re-form rhodopsin.

Both rod and cone visual pigments consist of 11-*cis*-retinal bound to a large protein, but they differ in terms of the wavelengths of light they optimally absorb. The reason is that the opsin proteins differ—rod opsin is different from any of the cone opsins, and the cone opsins differ among themselves. It is the interaction between the 11-*cis*-retinal and the opsin protein that determines the wavelengths of light a visual pigment molecule best absorbs. And because rod and cone opsin proteins differ, the rod and cone visual pigments absorb maximally different wavelengths of light. In humans, the rod visual pigment absorbs light maximally at 500 nm, in the green-blue region of the spectrum, whereas the three cone pigments absorb light maximally at about 560 nm (red/yellow light), 530 nm (green light), or 430 nm (blue light) (plate 4).

Because proteins are coded by genes, different genes code for the rod and cone opsins, as first demonstrated by Jeremy Nathans and his colleagues in the 1980s. Defects or alterations in these genes lead to significant visual abnormalities. Defects in the rod gene cause loss of rods and dim-light (night) vision. Many types of inherited retinal degenerative diseases such as retinitis pigmentosa (RP) are caused by mutations in the gene coding for rod opsin. Defects in or loss of one or more of the genes coding for the cone opsins lead to color blindness. Red-blind individuals are missing or have an altered gene for the red/yellow-sensitive visual pigment; green-blind individuals are usually missing the green opsin gene; and blue-blind individuals have an altered blue pigment gene. In addition to individuals who are missing one or another of the cone pigments because of mutations that prevent an opsin from forming, there are other individuals that have all three cone receptors, but one of the receptors has a cone pigment that does not absorb light exactly as does the normal pigment. These individuals are trichromatic, as are normal individuals, but they perceive colors somewhat differently from normals and are called anomalous trichromats, whereas those who are missing one cone type are called dichromats. Interestingly, human dichromats have a normal number of cones; the remaining cones fill in for the missing cones. As far as incidence is concerned, probably 80% of those who have color deficits are anomalous trichromats.

Those who are termed color blind because they are dichromats or anomalous trichromats can still distinguish colors. Only two cone types with different spectral sensitivities are required to discriminate colors. These individuals do not perceive certain hues as do normal individuals. So, for example, protanopes who are missing the red-sensitive cones, have trouble distinguishing reddish hues; deuteranopes green hues; and tritanopes blue hues. How much difficulty anomalous trichromats have in distinguishing hues depends on how different the spectral sensitivity of the anomalous pigment is from that of the normal visual pigment.

Interestingly, most mammals are dichromats, whereas fish and other nonmammalian species are often tetrachromatic—they have more than three cone types, each of which contains a different visual pigment. Most mammals are dichromats because mammals arose from shrew-like creatures during the age of the dinosaurs and were nocturnal. They had no need of all four of the visual pigments that had arisen during evolution up to that point. Thus, they lost two of the four and were dichromats with a short-wavelength, blue pigment and a medium-wavelength, green pigment. Not until the primates arose did trichromacy appear, and this resulted from a duplication of the medium-wavelength (green) pigment gene. This occurred about 35 million years ago, but to this day, the green- and red/yellow-sensitive visual pigments are quite similar in the wavelengths of light they maximally absorb. Indeed, these pigments differ by only about 30 nm in the wavelengths of light they maximally absorb, and this represents a difference between the two opsin proteins of only 15 amino acids. Thus, an alteration in just one amino acid in the red/yellow or green opsin proteins can make a difference in terms of the maximum wavelength of light these pigments absorb.

In humans, the genes for the red and green opsin proteins are on the X chromosome, one of the sex-determining chromosomes. Because males have one X chromosome and females two, red/green blindness is much more common in males than in females. In other words, only one X chromosome (one copy of the gene) is needed for normal color vision. Females have two X chromosomes, so one can be defective and color vision remains normal, whereas if the one X chromosome in a male is defective, that individual will be color blind, which is why red/green color blindness is described as sex-linked. The gene for the blue pigment is not on the X chromosome, and so the incidence of blue blindness is equal in males and females.

The visual cycle makes it clear why vitamin A is required for normal vision. Individuals deficient in vitamin A cannot make normal amounts of visual pigment and are, therefore, less sensitive to light, a condition known as night blindness because of the loss of light sensitivity by the rods. Cones also lose sensitivity in vitamin A deficiency, but because they operate in bright light, this sensitivity loss is seldom noticed. Today there is little nutritional night blindness in developed countries around the world, but vitamin A deficiency and nutritional night blindness remain a significant problem in many third-world countries.

Phototransduction

Light-activated visual pigment molecules lead to excitation of the photoreceptor cell through a cascade of biochemical reactions called phototransduction that were elucidated mainly in the 1980s. As shown in figure 3.4, rhodopsin (Rh*), when activated, first interacts with a molecule called transducin. (It is also called a G-protein because it binds a small, energy rich molecule called guanosine triphosphate [GTP]). Transducin, in turn, activates an enzyme called phosphodiesterase (PDE). This enzyme breaks down a small molecule, cyclic GMP (cyclic guanosine monophosphate), which in the dark is quite abundant in the outer segments. Cyclic GMP interacts with a membrane protein that forms channels in the membrane of the outer segment and that allows Na^+ and Ca^{2+} to enter the cell. When cyclic GMP is broken down by PDE, its levels fall, and the protein channels in the membrane close. This changes the voltage or potential across the cell's outer membrane, resulting in the inside of the cell becoming more negative. That is, the number of positive Na^+ and Ca^{2+} ions entering the cell is reduced, making the inside more negative relative to the outside.

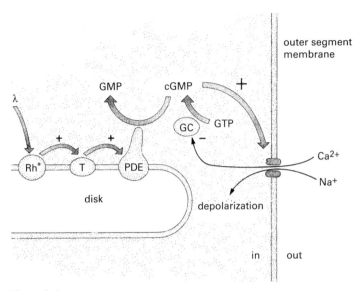

outer segment
membrane

GMP cGMP +

λ

GC GTP

−

Ca²⁺

Rh* T PDE

Na⁺

disk depolarization

in out

Figure 3.4

A simplified scheme of the phototransduction process in the photoreceptor outer segment. Light-activated rhodopsin (Rh*) activates transducin (T) a G-protein, which in turn activates the enzyme phophodiesterase (PDE). Phophodiesterase breaks down cyclic GMP to an inactive
product (GMP); in the absence of cGMP, which opens channels in the
outer-segment membrane, the channels close, and the cell hyperpolarizes (becomes more negative inside) because the positively charged Na^+
ions no longer can enter the cell. Also, Ca^{2+} levels in the cell decrease, allowing guanylyl cyclase (GC), normally inhibited by Ca^{2+}, to increase the
synthesis of cGMP from GTP. With more cGMP available, more channels
open, countering the effects of light. Ca^{2+} thus plays a role in photoreceptor adaptation.

Several additional points should be made here. First, one pho-
toactive visual pigment molecule can activate many transducin
molecules, and one PDE molecule can break down hundreds of
cGMP molecules. Thus, as a result of the activation of one visual
pigment molecule, many channels in the outer membrane of
the outer segment close. Thus, a significant amplification of the
light signal occurs during the phototransduction process.

Second, most cells in the nervous system open channels in
the cell membrane when they are stimulated. These channels
bring ions with a positive charge into the cell, causing the cell
to become more positive inside (called depolarization). Synaptic
transmission from one cell to the next is activated when the
transmitting cell becomes more positive. With vertebrate photoreceptor cells, the opposite happens. When the cell is illuminated, channels in the outer membrane close, causing the cell to
become more negative inside (hyperpolarization) and the cell's
synapses to become less active. In other words, photoreceptors
behave as though they are stimulated by darkness and that light
turns them off, an unusual situation and for which there is currently no explanation.

Third, Na^+ ions are much more abundant in the nervous system than are Ca^{2+} ions, and so most of the change of potential (voltage) that occurs in photoreceptors in light and dark is
the result of changes in Na^+ levels inside the cells. Photoreceptors are continually pumping Na^+ out of the cells in both light
and dark, but with the membrane channels mainly closed in
the light, the pumps do not need to work as hard, and so the
metabolic demand of the cells is significantly reduced. The high
metabolic demand of the photoreceptors, especially in the dark,
may have consequences for some retinal diseases, such as diabetic retinopathy.

Fourth, although Na^+ is the main ion regulating the potential
of photoreceptors in both light and dark, some Ca^{2+} ions also

go through the channels in the membrane, and they play an important role in regulating the response of the photoreceptor cell in the light. As indicated in figure 3.4, Ca^{2+} ions inhibit the enzyme guanylyl cyclase (GC), which makes cyclic GMP from GTP. In the light, when the channels that allow Ca^{2+} into the cell shut down, the inhibition of guanylyl cyclase by Ca^{2+} is relieved, and more cyclic GMP is synthesized, partially counteracting the breakdown of cyclic GMP by PDE. This results in the reopening of some channels in the outer-segment membrane. This phenomenon is called adaptation and is particularly important in bright light when all the outer-segment membrane channels have closed as a result of a substantial drop in cyclic GMP levels. The reopening of some of the channels allows the cell to continue to function regardless of light level.

Electrical Responses of Photoreceptors

Photoreceptors, in common with all neural cells, signal information by alterations in membrane voltage across the cell's outer membrane. At rest, neural cells generally have a steady resting voltage of between 60 and 70 millivolts (mV), with the inside of the cell negative to the outside. Put in other terms, there is an excess of negatively charged ions inside the cell at rest. When excited, channels in the cell's outer membrane open, allowing positively charged ions (usually Na^+) into the cell, causing the cell and membrane voltage to depolarize—that is, to become less negative. When a nerve cell is inhibited, channels in the membrane open that allow negatively charged ions (mainly Cl^-) to enter the cell, causing cell and membrane voltage to become more negative, toward the resting potential or beyond (hyperpolarization). Depolarization is linked with excitation because activation of synapses is caused by depolarization of a neuron, as noted above.

We have already learned that vertebrate photoreceptors are unusual in that darkness acts as the stimulus for them, whereas light turns them off. That is, in the dark, channels in the outer-segment membrane are open, allowing Na⁺ into the cell and causing the membrane to be depolarized (figure 3.4). In the dark the membrane potential is roughly –30 mV. The voltage difference between the inside and outside of a cell can be measured by inserting an electrode into a photoreceptor cell, as shown schematically in figure 3.5.

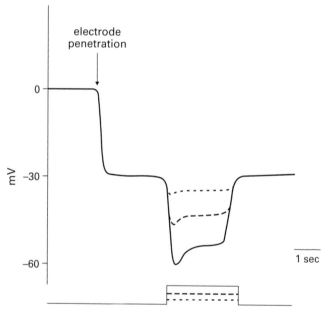

Figure 3.5
Idealized intracellularly recorded electrical responses from a vertebrate photoreceptor cell. The resting potential of this cell is indicated as –30 mV, and the cell hyperpolarizes (becomes more negative) to steps of light in a graded fashion. Note the relaxation of the response after the onset of light from a peak potential to a plateau. The duration and relative intensity of the light stimuli are shown at the bottom (dashed and straight lines).

When a light shines on a photoreceptor (as indicated on the bottom of figure 3.5), the channels in the outer-segment membrane close, causing the cell's membrane voltage to become more negative, the extent of which is intensity dependent as shown first by Tsuneo Tomita in the 1960s. If a dim light is used (short dashes), the response is small—only a few millivolts in amplitude. If the light is bright (continuous line), the membrane voltage may become more negative by as much as 30 mV or to the resting level of most neurons at rest, indicating that all or virtually all of the Na^+ channels in the outer-segment membrane are now closed.

But because lower Ca^{2+} levels raise cyclic GMP levels in the photoreceptor cell, membrane potential partially recovers. Again this is critical when a photoreceptor is in bright light because without such adaptation the cell would no longer be able to signal—all the membrane channels would be closed. Indeed, in rods this happens in bright light—they no longer can function, a phenomenon called rod saturation. Cones, fortunately, never saturate regardless of light levels.

There are other differences between rod and cone function. Rods that are dark adapted are about 25 times more sensitive than dark-adapted cones. However, cone responses tend to be faster than rod responses, and after light exposure, they recover more quickly.

Light and Dark Adaptation

We have described that, in continuous light, the response of photoreceptors is not maintained but recovers partially. There are other changes that occur in the photoreceptors when in light. The most striking is that they lose substantial light sensitivity. This loss of light sensitivity is not simply because levels of visual pigment decrease in the light, making the photoreceptor a less effective photodetector; rather, the sensitivity loss is far greater. So, for example, if in a rod half the rhodopsin is inactivated in the light, the sensitivity of the rod should be reduced by half or two times, but with half the pigment bleached, the sensitivity of a rod declines by 1000 to 10,000 times. Figure 3.6, (left side), shows schematically the loss of sensitivity that occurs in both the rod and cone systems as a function of light. As light intensity is increased (in logarithmic steps—by 10 times for each step), the threshold (inverse of sensitivity) increases logarithmically. In dim background light, visual sensitivity is governed by the rod system, the cone system does not come into play until the background light is about 2 log units (100 times) above absolute threshold. Then the cone system is more sensitive than the rod system, but its threshold also rises logarithmically as the light is increased further.

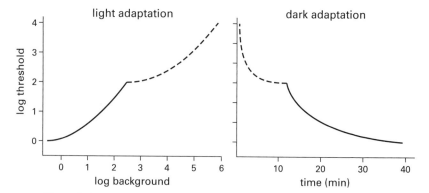

Figure 3.6

Schematic diagrams of typical light and dark adaptation experiments in the human. (Left) Light adaptation: the rise of visual threshold is plotted as a function of background light intensity. (Right) Dark adaptation: the recovery of visual threshold following bright-light adaptation is plotted as a function of time in the dark. In both kinds of experiment, two segments are observed. The upper segments (dashed lines) reflect cone adaptation; the lower segments (continuous lines), rod adaptation. See text.

When little visual pigment is bleached in a photoreceptor by an adapting light, subsequent recovery (called dark adaptation) is quite rapid, requiring only seconds to minutes at most. When much of the pigment in the photoreceptors is bleached by a bright light, it takes 30–40 minutes or more for the sensitivity to recover fully to dark-adapted levels. This is shown on the right side of figure 3.6. The cone system recovers sensitivity first, requiring 5–10 minutes, and then the rod system recovers, requiring another 20–30 minutes.

This phenomenon is evident when we enter a dark theater on a bright, sunny afternoon. Initially we see little and stay in the back of the theater for 5–10 minutes until seats become visible. After another 20–30 minutes or so, everything in the darkened theater is quite visible. What we are waiting for is the visual pigments in the cones and rods to re-form as was first shown by one of us (J.E.D.) in the United States and William Rushton in England in the early 1960s. Cone pigments regenerate faster than the rod pigments, explaining why we recover cone sensitivity before rod sensitivity.

Adaptive mechanisms also occur in the retinal synaptic network and in higher visual centers, but the major changes in sensitivity experienced when going from light to dark and back again reflect mainly changes occurring in the photoreceptors.

Inherited Retinal Degenerations—Retinitis Pigmentosa

Retinitis pigmentosa (RP) refers to a group of inherited diseases characterized by photoreceptor cell degeneration. Several hundred mutations in more than 100 different genes have been identified that cause RP, and there are probably many more. Fortunately, RP is rare—about one person in 3500 is affected. RP is usually first manifest in young adults and is more frequent and more severe in males. The rods are first compromised and the initial symptoms are diminished night vision and loss of peripheral vision. Ultimately the cones are affected as well, with progressive loss of central visual acuity and color perception. Many RP patients are legally blind by the age of 40, and the ultimate outcome is complete loss of all vision. One form of the disease, Usher syndrome, results not only in vision loss but hearing loss as well.

The genetics of retinitis pigmentosa are varied and complex. Often, RP is inherited dominantly; that is, a mutation in just one gene leads to the disease. The children of a parent with the defective gene have a 50% chance of inheriting the disease. RP may also be inherited recessively, in which the mutation must be in the same gene in both parents. These parents are known as carriers of the mutated gene, but they are not affected. On average 25% of their children will have the disease.

Of the many gene mutations that cause retinitis pigmentosa, almost 100 occur in the gene that codes for the rod opsin protein, as first shown by Ted Dryja and Eliot Berson in the early 1990s. Figure 3.7 shows a schematic drawing of rod opsin and where many of the known mutations lie in the chain of amino acids that makes up the protein. Each circle represents a different amino acid, and a filled circle means if this amino acid is changed as a result of a mutation in the opsin gene, RP will ensue, and one amino acid can be changed to several different amino acids depending on the mutation. Such amino acid mutations are found all over the molecule. There are also differences in the disease depending on where the mutated amino acid resides. Mutations on the right side of the molecule (near the C-terminal end) cause an earlier onset and a faster disease progression than do mutations near the N-terminal (left side of the molecule). Why the rod opsin gene is so susceptible to mutations is not known. One might expect that the cone opsin genes would be equally vulnerable, but this appears not to be the case.

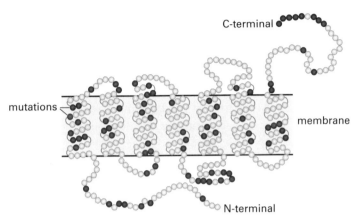

Figure 3.7
A schematic drawing of the rhodopsin molecule. A number of the amino acids that, when altered as a result of a genetic mutation, cause retinitis pigmentosa are indicated by black circles. C-terminal, carboxy terminal; N-terminal, amino terminal.

There are some inherited cone degenerations, but nowhere near the large number of disease-causing mutations that occur in the rod opsin gene.

There are mutations in other genes that code for various proteins in the rods or pigment epithelium, including transducin, PDE, and the isomerase that converts all-*trans*-vitamin A to 11-*cis*-vitamin A. Other mutations also affect critical photoreceptor/pigment epithelial cell proteins and lead to photoreceptor degeneration. There remain, however, many puzzles with regard to the mutations and the character of the disease they cause. For example, two individuals with identical mutations and even from the same family can have different onset times for the disease and different progression rates, indicating that other factors are involved, including, perhaps, environmental factors such as light exposure or other genetic factors (mutations in other so-called modifying genes). A dramatic example is that of two sisters, a year apart in age, with the same mutation. When they were in their late 40s the younger sister was virtually blind, whereas the other could still see well enough to drive her truck.

Retinitis pigmentosa and other inherited retinal degenerations have up to the present not been treatable. However, two recent procedures are promising for at least some cases and are in clinical trials. One is to inject a trophic or growth factor into the diseased eye to slow down the disease progression. In animals with inherited retinal degenerations or in which the photoreceptors were light damaged, this procedure slowed progression of cell death significantly, as shown by Matthew LaVail and his colleagues in the early 1990s. Although this is not a cure for a disease, delay of blindness by several years in those afflicted with retinitis pigmentosa would be of significant benefit.

The second and more promising procedure involves gene therapy—replacement of the defective gene—which has the potential to result in a permanent cure. Recently, gene therapy has been used by Jean Bennett and colleagues on a group of human patients with Leber's congenital amaurosis, a previously untreatable degenerative disease that causes severe loss of vision. In a recent research study 12 patients, ages 8 to 44, with Leber's disease received injections into one eye of the normal gene using a virus vector. The virus infected the pigment epithelial cells where the injected normal gene began to express protein. The gene involved codes for the protein mediating the isomerization of vitamin A in the pigment epithelium, which in its mutant form makes very little 11-*cis*-vitamin A. All of the participants demonstrated significantly improved vision, particularly the children, with no significant complications, demonstrating that the injected gene was now making abundant isomerizing protein. This is an exciting advance that has the potential to benefit many more types of retinal degenerations.

Whereas gene therapy has been touted for many years as a way to cure gene-mutation diseases, progress has been slow. The difficulty in applying this technique more widely to retinitis pigmentosa and other inherited retinal diseases results from the extremely large number of mutated genes associated with inherited retinal diseases. A different viral construct is necessary for each mutated gene. Another basic research procedure under study is to inject into the eye a gene coding for a light-sensitive molecule. When coupled with a gene that promotes expression of genes in a specific retinal cell type (a promoter gene), it has been possible to make ganglion, bipolar, and even some remaining photoreceptors light sensitive. Studies in animals have shown that this technique does restore some vision. In chapter 7, we describe in more detail these and other approaches to cure retinal degenerations.

Age-Related Macular Degeneration

The macula is a small area in the center of the retina that contains the fovea and mediates sharp vision (figure 2.1). The fovea is a small (0.5-mm) indented area that in its center possesses only cone photoreceptors that are the thinnest and longest of all retinal photoreceptors (figure 1.5). As we have already noted, a normal functioning fovea is necessary for high-acuity visual tasks such as recognizing faces, reading fine print, watching television, and driving. There are only about 35,000 cones in the fovea out of 6 million cones in the human eye, and the inner layers of the retina are displaced laterally around the fovea, which allows rays of light to have more direct access to the photoreceptors (see figure 1.5). Further, there are no blood vessels impinging on the fovea, which could distort images falling on the cones. Blue-sensitive cones are also mainly excluded from the fovea because blue light is refracted (bent) more than are red and green lights by the cornea and lens, and, therefore, blue light images focus at a different level than do red and green lights. The fovea is also protected from blue light by the presence of yellow pigment in the macula that strongly absorbs blue light, preventing it from reaching the photoreceptors.

As with all structures in the human body the macula and fovea are subject to a variety of abnormalities and diseases, especially those associated with the aging process. The most common and dreaded disease today is age-related macular degeneration (AMD), which is the most frequent cause of poor visual acuity in older (over 60) Americans. Although rare, forms of macular degeneration do occur in children and young adults. Currently some 9.1 million Americans have documented macular degeneration, and it is anticipated that this number will increase as average life expectancy increases.

AMD usually has a gradual onset. The first symptom is mildly blurred vision, which progressively increases to the point where objects such as faces or road signs are no longer sharp and clear. A prominent early symptom is that straight lines appear to have a wavy defect in their center.

AMD occurs in two forms, dry (atrophic or nonvascular) and wet (neovascular). An ophthalmologist can often recognize an eye at risk of AMD before any visual changes have occurred by the presence of small deposits under the retina called drusen. Dry AMD, the more benign of the two forms, is consistent with reasonably good vision for a long time. In most instances dry AMD begins with a patchy loss of both cones and rods in areas around the macula but then can extend into the macula and fovea. This stage, known as geographic atrophy, can result in severe central visual loss. About 15% of cases of dry AMD convert to the more severe wet form, which is characterized by the growth of new and defective blood vessels into the macula region of the retina. These defective blood vessels leak fluid and blood, often causing a sudden severe hemorrhage with permanent central visual loss and a scar forming over the site of the destroyed macula. Growth of these defective blood vessels is stimulated by a protein called vascular endothelial growth factor (VEGF).

Plates 5, 6, and 7 are photographs of the appearance of the inside (the fundus) of a normal eye (top) as seen by an ophthalmologist examining the eye with an ophthalmoscope and eyes with early and late stages of AMD (below). Note the orange-red color of the fundus in all photographs. When one looks at the pupil of an eye, one sees only a black reflex. This is because the viewer's head is blocking the light rays that would ordinarily enter the eye and reflect back the normal red reflex as well as the structural features of the fundus. An ophthalmoscope has a light that enters the eye unblocked and reflects back the fundus color. This is the same reason photographs taken with a flash show the annoying red pupil reflex.

Note the round optic disk in the normal fundus (plate 5) composed of the ganglion cell fibers on their way to the brain. Also seen are the smaller, bright red arterial blood vessels arising from the central retinal artery in the middle of the optic nerve and the parallel darker blood vessels, the retinal veins, bringing blood back to the general circulation. To the right of the optic nerve is a darker area without blood vessels, the macula, and in its center is the fovea.

In the eye at risk for AMD (plate 6), drusen deposits are seen. In the fundus of the eye with late-stage AMD (plate 7), the macula is essentially obliterated by a retinal scar. Clearly this patient has essentially no foveal vision in this eye, and the eye would be characterized as legally blind.

Over the past few years we have gained some insights in the underlying cause of AMD. It starts in the rod-rich area surrounding the fovea. Figure 3.8 shows the distribution of rods and cones in the human retina. In the central fovea itself there are

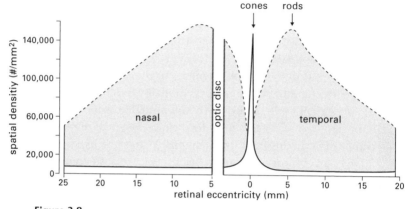

Figure 3.8

The distribution of rods and cones across the retina. The fovea (0 eccentricity) contains only cones. The highest density of rods surrounds the fovea. There are no photoreceptors in the optic disk, where the ganglion cell axons exit the eye.

no rods but a very high density of cones. Adjacent to the fovea is a high rod density, which gradually decreases as the distance from the fovea increases. The number of cones, on the other hand, is quite constant across the entirety of the retina once one leaves the foveal region.

One theory as to what happens in the rods to cause AMD is the following. When visual pigment molecules are bleached, retinal separates from the protein opsin. Retinal, an aldehyde, is a highly reactive molecule, and when it is released, it is usually sequestered by opsin or other proteins in the outer segments until the retinal can be converted to vitamin A, which is not very reactive. If some retinal escapes, it is free in the outer-segment membranes and can then react with a membrane lipid, which initiates the formation of a very stable and toxic compound called A2E.

Intermediate compounds involving retinal and the molecules it reacts with move into the pigment epithelium, where the final formation of A2E occurs. In the pigment epithelium, A2E damages the lysosomes, intracellular organelles responsible for digesting cellular debris including the disks shed from the photoreceptors. With failure of the lysosomes, substances accumulate in the pigment epithelium, gradually poisoning the cells. The pigment epithelial cells may secrete some of this material, contributing to drusen formation. With poisoning of the pigment epithelium, it no longer can provide 11-*cis*-retinal to the outer segments nor digest the outer segment material normally shed by the photoreceptors. Debris accumulates, and the photoreceptors adjacent to these pigment epithelial cells begin to die, leading to the patchy lesions of dry AMD.

Relatively little retinal ordinarily escapes into the outer segment membranes. Thus, it takes years for enough A2E to accumulate to damage the pigment epithelium. Other compounds, called *bis*-retinoids, also form from retinal, and they add to the toxicity. Dry AMD transitions to wet AMD by an inflammatory reaction, causing blood vessels to grow into the fovea. A clear risk factor for AMD, especially wet AMD, is smoking, and the presumption is that smoking leads to the oxidation of the toxic retinal-based compounds. Oxidation of these compounds clearly enhances the inflammatory reaction, and, in particular, it strongly promotes the release of VEGF, a molecule responsible for initiating abnormal blood vessel growth.

There is today no effective treatment available for dry AMD, although many scientists are working to discover a cure. The recent Age-Related Eye Disease Studies (AREDS) sponsored by the National Eye Institute demonstrated that a combination of high doses of vitamins A, C, and E, which act as antioxidants when combined with zinc and copper, slow the progression of dry AMD by about 25%, although this regimen does not appear to prevent the disease. It has also been demonstrated that people eating a diet high in green leafy vegetables and fatty fish such as salmon have a lower incidence of AMD. So a high-dose vitamin and mineral combination and frequent consumption of the two food groups noted above are recommended to all patients who have early AMD.

There is treatment available for the wet (neovascular) AMD, which consists of injections of anti-VEGF compounds directly into the vitreous humor. These injections prevent the development and enhancement of new and defective blood vessels in and around the macula. This treatment, which must be continued almost indefinitely, can slow significantly the progression of the disease and in many instances results in some improvement of vision.

The diagnosis of AMD (and other retinal diseases) was classically made by direct observation using an ophthalmoscope (plates 6 and 7). Optical coherence tomography (OCT), a recent development in imaging technology, first appeared in the scientific literature in 1991. Since then it has revolutionized the diagnosis and management of eye diseases, particularly AMD and diabetic retinopathy. The technology has advanced spectacularly since it was introduced, and today it provides detailed views of both normal and diseased retinas. It provides an image of the retina similar to a histological section, such that one can observe the three layers of cells, the two plexiform layers, and even the inner and outer segments of the photoreceptors. In the not-too-distant future it may very well replace the ophthalmoscope, the instrument used to examine the retina in living humans since it was introduced by Hermann von Helmholtz in the nineteenth century.

A

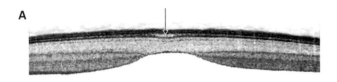

B

250 μm

Figure 3.9
(A) An optical coherence tomography (OCT) image of a normal fovea and surrounding retinal area. The clear area above the fovea (arrow) represents the elongated foveal cones (see figure 1.6). (B) An OCT image from a patient with wet AMD. The foveal and nearby photoreceptors are pretty much destroyed (right), and there is an accumulation of fluid on the left that has caused the retina to detach from the pigment epithelium.

Figure 3.9 shows an optical coherence tomographic (OCT) image of a normal eye through the fovea (A) and eye with wet AMD (B). In the eye with wet AMD, the central cones have been lost, and the retina to the left has been detached from the pigment epithelium because of the accumulation of fluid (edema) from defective blood vessels that have grown nearby. This patient likely has little central vision remaining.

4 Analyzing the Visual Image—The Retina:

Glaucoma and Diabetic Retinopathy

Two stages of visual information processing occur in the retina—one in the outer plexiform layer, the other in the inner plexiform layer (see figure 1.8). Both plexiform layers are built on a similar plan. An input neuron brings visual information into the plexiform layer and synapses onto an output neuron, which transmits the visual message from the plexiform layer. In addition, there are interneurons in each plexiform layer that make synapses with both the input and output neurons, thus modifying the visual signals in each plexiform layer.

The photoreceptors provide input to the outer plexiform layer (OPL), whereas the bipolar cells are the output neurons for the OPL (figure 1.8B). The horizontal cells serve as the OPL interneurons, providing for lateral interactions between the photoreceptor cells and between the photoreceptor cells and bipolar cells. The inner plexiform layer (IPL) is organized similarly; bipolar cells provide the input to the IPL, ganglion cells are the output neurons, and amacrine cells are the interneurons. The amacrine cells serve to modify and/or extract features from the input of the bipolar cells and encode them in the responses of the output ganglion cells.

There are numerous subtypes of the various retinal neurons, so many synaptic interactions occur in each plexiform layer.

This results in subtypes of neurons displaying different response properties to visual stimuli. For example, certain of the output neurons of the retina (the ganglion cells) are concerned with the distribution of light on the retina—they code spatial information; other ganglion cells respond best to movement and, in some animals, even the direction of movement—they code temporal information; and finally, some ganglion cells respond selectively to different wavelengths of light—they code color information.

A good deal of present retinal research is concerned with elucidating the synaptic interactions that result in the various response properties of the neurons, and there is still much to learn. Neurons in the retina, like neurons throughout the nervous system, communicate synaptically mainly by chemical means. At an activated synapse, a chemical is released that diffuses across a narrow cleft to interact with proteins on the membrane of the contacted (postsynaptic) cell. This results in the excitation, inhibition, or modulation of the postsynaptic neuron.

When a neuron is synaptically excited, the activated proteins form channels in the cell membrane, allowing positively charged ions to flow into the cell, and the neuron depolarizes—it becomes more positive inside (figure 4.1). At an inhibitory synapse the protein channels admit negatively charged ions into the cell, and the cell hyperpolarizes. Modulation results when postsynaptic membrane proteins are activated by substances released at a synapse and interact with intracellular enzyme systems via G-proteins (such as transducin in phototransduction—see chapter 3), thereby modifying the cell biochemically. Neuromodulation can alter various properties of neurons and their synapses, thus altering synaptic strength and neural circuit effectiveness. There are also electrical synapses in the retina, where connecting channels between two neurons allow ions to flow directly from one cell to another.

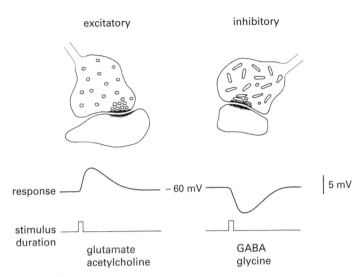

Figure 4.1
Typical excitatory and inhibitory synapses, as observed by electron microscopy. Excitatory synapses are characterized by round synaptic vesicles; inhibitory synapses often have flattened vesicles. The chemicals released at chemical synapses (called transmitters) are stored in the synaptic vesicles. Membrane specializations, along with a cluster of vesicles, are characteristically seen at sites of transmitter release. At excitatory synapses small depolarizing (positive) potentials are elicited in the postsynaptic cell; at most inhibitory synapses, hyperpolarizing (negative) potentials are generated, thereby making the membrane potential of the neuron more negative. Glutamate and acetylcholine are typical excitatory transmitters; γ-aminobutyric acid (GABA) and glycine are inhibitory transmitters.

In this chapter we consider in more detail the neurons contributing to each plexiform layer, the synapses they form and that are formed on them, and their responses. We describe in detail two ganglion cell subtypes whose response properties are quite well understood in terms of the synaptic interactions giving rise to their responses. Toward the end of the chapter we discuss two common and serious diseases, glaucoma and diabetic retinopathy; glaucoma primarily affects the retinal ganglion cells and the retinal output, whereas diabetic retinopathy likely relates to the unusual properties of the photoreceptors and many of the retinal neurons that are stimulated by darkness and turned off by light.

Outer Plexiform Layer—Form and Color

The photoreceptor cells make synapses onto the bipolar and horizontal cells in a complicated way. What is known is that photoreceptor terminals provide input to two basic types of bipolar cells—one that depolarizes in response to light, termed an ON-bipolar cell—and one that hyperpolarizes in response to light, an OFF-cell (figure 4.2A). Why are both ON and OFF bipolar cells needed? Experiments in the mid-1980s by Peter Schiller and his colleagues, in which the ON-bipolar cells in a monkey were prevented from responding by pharmacological means, showed that such animals could no longer tell if a spot of light was becoming brighter than the background, but they could still tell if the light was becoming dimmer than the background. These data suggested that ON-cells signal illumination increases, whereas OFF-cells signal illumination decreases.

Horizontal cells mainly hyperpolarize to illumination, much as the photoreceptors do, as first observed by Gunnar Svaetichin and Edward MacNichol in the late 1950s. The synapse between photoreceptor and horizontal cells acts as an excitatory synapse.

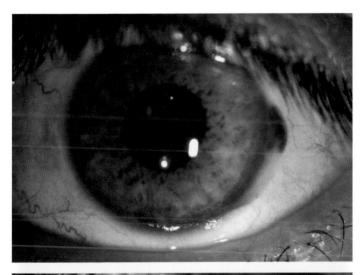

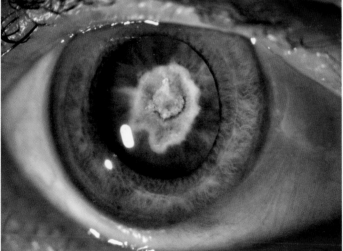

Plate 1

The normal human eye. The black area in the center is the pupil, whose diameter is controlled by the iris surrounding the pupil. Eye color is determined by pigments in the iris.

Plate 2

An advanced cataract with a dense nuclear opacity and a surrounding area with spoke-like opacities that extend out across much of the lens.

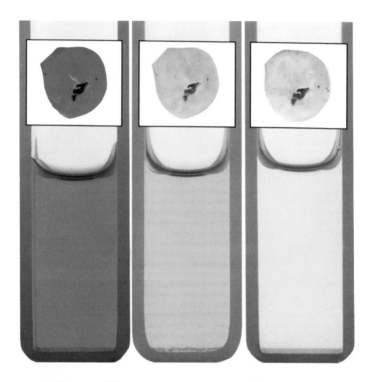

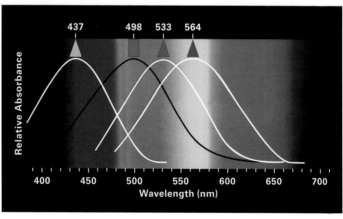

Plate 3

Retinas from the bullfrog, along with extracts of retinas, showing the color of the rod visual pigment rhodopsin (left images) and the color changes that take place on exposure to light (right images). Initially on light exposure, the retinas and extracts turn yellow; then the yellow color gradually disappears, and the retina is rendered white and the extract colorless. Rhodopsin consists of retinal bound to the protein opsin, but light causes the retinal to separate from the protein, and the yellow color is due to the free retinal. With time, the retinal gets converted to vitamin A, which is colorless.

Plate 4

Absorption spectra of the three cone pigments and the rod pigment rhodopsin from measurements made on human and monkey retinas. The numbers at the top indicate the peak absorptions of the pigments.

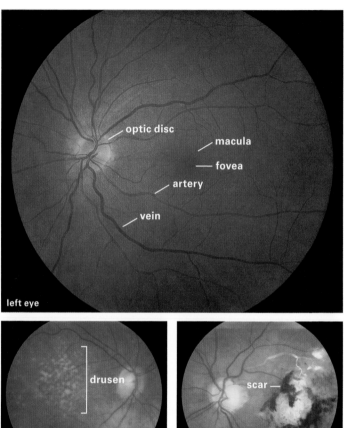

optic disc

macula

fovea

artery

vein

left eye

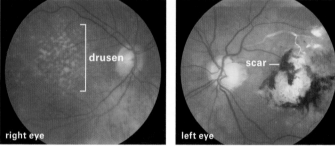

drusen

right eye

scar

left eye

Plate 5

The normal fundus (back of the eye) as viewed by an ophthalmoscope. The optic disk is where the ganglion cells axons exit the eye and is known as the blind spot because there are no photoreceptors in this area. Blood vessels that nourish the inner retina enter the eye via the optic nerve. The macula is the darker area to the right and is free of blood vessels. The fovea is in the center of the macula.

Plate 6

An eye showing drusen deposits in the central retina. Drusen deposits suggest the eye is at risk for AMD.

Plate 7

An eye with late-stage wet AMD with a large scar that essentially obliterates the macula and fovea. Visual acuity of this eye would be less than 20/200.

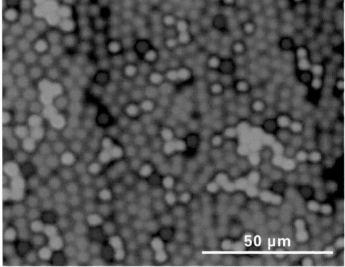

50 μm

Plate 8

A depiction of how the colors we perceive depend on background color, inspired by a painting by Joseph Albers. The colors of the X's on both sides of the figure are identical, but on the left the X appears quite blue, whereas on the right it appears yellow. If you look at where the X's come together, you can see they are identical in color.

Plate 9

The arrangement of cones in the human fovea as demonstrated by adaptive optics. There are many more red than green cones in most human retinas, and fewer blue cones, especially in the fovea. Indeed, the very center of the fovea is devoid of blue cones.

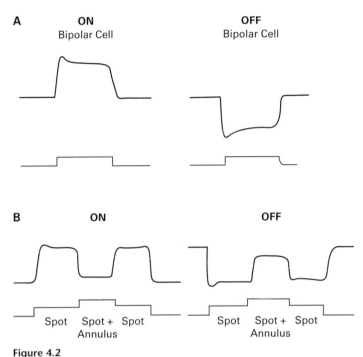

Figure 4.2

Responses of bipolar cells to light. (A) About half of the bipolar cells depolarize to light and are designated ON-cells; half hyperpolarize and are OFF-cells. (B) Both ON- and OFF-bipolar cells show surround antagonism. When an annulus (ring of light) is added to a spot of light, the cell's response is diminished, even though more light falls on the retina.

In the dark, the synapses are active, resulting in depolarization of the horizontal cells (see chapter 3 about photoreceptor depolarization in the dark); when the photoreceptors hyperpolarize in the light, the synapses turn off, thereby hyperpolarizing the horizontal cells. OFF-bipolar cells behave much the same way—they are depolarized in the dark and hyperpolarize in the light. ON-bipolar cells, on the other hand, behave as if they are receiving inhibitory input from the photoreceptors—they depolarize in the light. The underlying mechanism is unusual. The proteins in the bipolar cell membrane that are activated by the neurotransmitter released by the photoreceptors in the dark interact with a G-protein to close channels in the bipolar cell membrane. In the light when the synapse is turned off, the channels open, allowing Na^+ to enter the cell, depolarizing the bipolar cell.

Horizontal cells play an important role via their processes that extend widely in the outer plexiform layer (see figure 1.8) and enable distant photoreceptors to influence bipolar cell responses. They provide lateral inhibition to both photoreceptors and bipolar cells such that bipolar cells display a center-surround receptive field organization (figure 4.2B). The receptive field of a neuron in the visual system is defined as the area on the retina that, when stimulated, alters the response of the neuron. All neurons in the retina and elsewhere along the visual pathways have receptive fields that respond to different stimuli—light or dark spots, colored or moving spots or lines, or, in the cortex, lines or edges that have a specific orientation. In the case of bipolar cells, the cells respond with either an ON or OFF response to illumination of the photoreceptors that they directly contact, that is, receptors just above them. A spot of light covering all of those receptors best drives the cell and generates the largest response. Distant photoreceptors, on the other hand, induce a response of the opposite sign in the bipolar cell, a surround response: ON-center cells have an OFF-surround response, whereas OFF-center cells have an ON-surround, first observed by Frank Werblin in the late 1960s. The center and surround responses inhibit each other so that, if both center and surround are illuminated simultaneously, only a weak response is elicited in the cell. Thus, the bipolar cell responds best when the light is restricted to the center or to the surround of the cell's receptive field. In other words, bipolar cells are interested in the spatial distribution of light on the photoreceptor mosaic. Further,

Figure 4.3
Both circular disks reflect the same amount of light, yet the difference in background illumination makes the right disk appear darker. From *Optical Illusions, Amazing Deceptive Images—Where Seeing is Believing* by Inga Menkoff. Created by Any Puzzle Media, published by Parragon Books Ltd, copyright © 2007.

because the center response is weakened by surround illumination, spots of light look brighter against a dark background than a bright background, a phenomenon we normally experience, as shown in figure 4.3.

How horizontal cells exert their effects is still largely unknown. They clearly provide feedback inhibition onto photoreceptors, but feedforward inhibition onto bipolar cells can also occur, and conventional synapses have been seen between horizontal and bipolar cells. Another interesting aspect of horizontal cells is that they are coupled to one another by electrical synapses. These electrical synapses serve to extend their lateral influence, which means the surround inhibition is exerted by multiple horizontal cells and is considerably larger than the lateral extent of a single horizontal cell.

Color processing also begins at the level of the bipolar cells, and the receptive fields of visual cells responding to color typically show opponent responses. Ewald Hering, back in the late nineteenth century, first proposed that opponency was a critical feature of color processing by pointing out that certain colors mix, giving different hues—red and blue yield purple, red and yellow give orange—but other colors do not mix—green and red, blue and yellow. Opponent processes could explain this, he reasoned, and suggested that red and green are opponent, as well as blue and yellow. There are, of course, no cone receptors that respond best to yellow light, but together the red and green cones respond maximally to yellow (see below).

In most color-opponent bipolar cells in mammals, the receptive field center responds best to light of one color (e.g., red), whereas the surround responds best to light of another color (e.g., green) or vice versa (figure 4.4). These are called single opponent cells. In fish and other nonmammalian species with good color vision, there are also double opponent bipolar cells in the retina. These cells have center and surround responses that change polarity with wavelength—that is, red light depolarizes the cell centrally, whereas green light hyperpolarizes the cell centrally or vice-versa. In the surround the red and green lights elicit responses opposite to the central responses. In primates there are also double opponent cells, but they are found only in the cortex (see chapter 5).

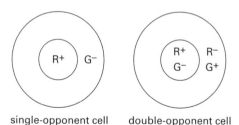

single-opponent cell   double-opponent cell

Figure 4.4
Receptive fields of single- and double-opponent color cells. R^+ means that red light excites that region of the receptive field; G^- means that green light inhibits that region of the field.

It has been pointed out that single opponent cells can explain color contrast; that is, following sustained gaze at a red piece of paper, a subsequent look at a white piece of paper results in a greenish appearance of the paper (and vice versa). Double opponent cells, on the other hand, could explain simultaneous color contrast; a gray field surrounded by a red field looks greenish, and vice versa.

Inner Plexiform Layer—Movement

The bipolar cells provide the input to the inner plexiform layer and synapse onto amacrine and ganglion cells. The inner plexiform layer is far thicker than the outer plexiform layer, and its synaptic interactions are much more numerous and complicated. As noted earlier, there are at least 12–15 different types of ganglion cells, the retinal output neurons, with regard to the type of information they carry, and each subtype receives its input on a different level in the inner plexiform layer as first demonstrated by Botand Roska and Frank Werblin in 2001. Many of these cell types project to the lateral geniculate nucleus and cortex, leading to visual perception, but others go to brain structures involved with other visual phenomena such as regulating pupil size, controlling eye movements, stabilizing the image on the retina, and modulating circadian rhythms.

An important feature of the neurons in the inner plexiform layer is their response properties. All of the retinal neurons we have discussed so far, photoreceptor, horizontal, and bipolar cells, respond to light with sustained, graded potentials. That is, the amplitude of their electrical responses varies with stimulus intensity—the stronger the stimulus, the greater the response, and their responses are sustained—the responses last for the duration of the time the light is on or off, although with both photoreceptors and bipolar cells, the initial response to light is greater than the response thereafter. Further, the size of the initial transient response varies with cell subtype. However, amacrine and ganglion cells generate a second type of electrical response, one that is found generally throughout the nervous system, called action potentials or spikes; they are large (>100 mV) and transient. Furthermore they are propagated, which means they do not lose amplitude as they travel along a cell. But because action potentials are transient, lasting only a few milliseconds, and are constant in size, they cannot code stimulus intensity information by amplitude of response. Rather, they code information by frequency of response; the more action potentials generated per unit of time, the more excited is the cell. Generation of action potentials by a cell depends on an underlying depolarization of the cell; that is, more depolarization means more spikes generated per unit of time.

Action potentials are used to carry neural information long distances in the nervous system—distances of more than about 1 mm. Distal retinal neurons do not generate action potentials because they are transmitting information over distances much less than 1 mm and have no need for a long-distance communication mechanism. However, some amacrine cells extend processes a millimeter or more within the inner plexiform layer, and ganglion cells must transmit the visual signal from eye to brain, a distance in most animals much greater than 1 mm. In summary, many amacrine cells and all ganglion cells generate action potentials.

Previous discussion suggested that there are a limited number of bipolar cells—ON- and OFF- center cells and color opponent cells—that provide the input to the inner plexiform layer. However, we now know that 10–12 different bipolar cell types provide input to the inner plexiform layer cells. They differ anatomically mainly in where their axons terminate—some terminate more distally in the inner plexiform, others more proximally, and

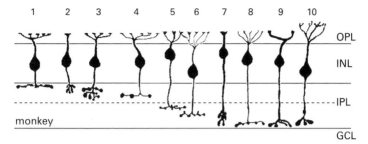

Figure 4.5

Drawings of bipolar cells of the monkey retina. The horizontal dashed line dividing the inner plexiform layer represents the border between the OFF- and ON-sublaminae. Bipolar cells whose axons terminate above that line are OFF-bipolar cells; those whose axons terminate below it are ON-bipolar cells. Note that within the OFF- and ON-laminae the terminals of the various bipolar cells end on one or another sublayer.

others in the middle (figure 4.5). We now appreciate that there may be as many as 10–12 layers or strata in the inner plexiform layer, and it is along these strata that the 12–15 different ganglion cells receive their input from bipolar and amacrine cells. How do the 10–12 bipolar cells differ physiologically? First, the OFF-bipolar cells terminate in the upper half of the inner plexiform layer, whereas the ON-bipolar cells all terminate in the lower half of the inner plexiform layer, first described by Edward Famiglietti and Helga Kolb in 1976. Bipolar cell terminals that receive mainly (or exclusively) rod input are found very deep in the inner plexiform layer. There are also some differences in the temporal responses of the various bipolar cells—some have more prominent initial transient responses than do others.

Many amacrine cells are highly stratified. Stratified amacrine cells as shown in figure 4.6, a drawing made by Ramón y Cajal in the late nineteenth century. Cajal, a Spaniard, described the cells in many structures of the brain including the retina and is

Figure 4.6
Drawings of amacrine cells by Cajal. Five stratified amacrine cells (right)
and a diffuse amacrine cell (left).

often called the father of neuroscience. Stratified amacrine cells
extend along one or another of the layers in the IPL and, presumably, are involved in generating the response properties of
the ganglion cells receiving input in that layer. Other (smaller)
amacrine cells extend their processes throughout the inner
plexiform layer and provide for synaptic interactions among the
layers. The wide-field stratified cells generally use one transmitter, γ-aminobutyric acid (GABA), whereas the narrow-field diffuse amacrine cells employ glycine as their neurotransmitter.
Both GABA and glycine exert inhibitory effects indicating that
amacrine cells are generally inhibitory interneurons like the horizontal cells in the outer plexiform layer.

Many amacrine cells (but not all) respond transiently at both
the onset and offset of illumination when the retina is illuminated with a static spot of light. However, if the spot is moved
around, the cell continually responds (figure 4.7). Thus, these
amacrine cells clearly play a role in the detection of movement,
and they impart this property to movement-sensitive ganglion
cells. Amacrine cells that respond in a more sustained fashion
likely play a role in ganglion cells that give sustained responses.

Ganglion cells may be stratified or diffuse but are mostly
stratified. There are also bistratified ganglion cells. A stratified
ganglion cell that extends processes in the upper part of the
inner plexiform layer (OFF-laminae) is an OFF-ganglion cell; if

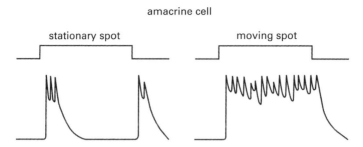

Figure 4.7
Idealized transient amacrine cell responses. To a stationary spot, the cells respond with transient responses at both the onset and offset of the light. To a moving spot, the cell responds continuously.

it extends processes in the proximal part of the inner plexiform layer, it is an ON-cell. Bistratified ganglion cells that extend processes in both the ON-and OFF-sublaminae are ON-OFF cells; to static spots of light, they respond at both the onset and offset of illumination.

ON-Center and OFF-Center Ganglion Cells

Many ganglion cells have receptive fields much like the ON- and OFF-center bipolar cells, but with some additional features. For example, they receive some inhibitory input from amacrine cells. ON- and OFF-center ganglion cells with inhibitory surrounds were first described by Stephen Kuffler in the United States and Horace Barlow in England in the early 1950s. These ON- and OFF-center cells are the most common ganglion cells found in mammals. Idealized responses and receptive field maps for such cells are shown in figure 4.8. When the centers of their receptive fields are stimulated with spots of light, they give a vigorous ON or OFF response, either a continuous burst of action potentials

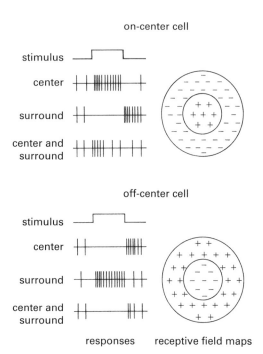

responses   receptive field maps

Figure 4.8

Idealized ganglion cell responses and receptive field maps for ON-center and OFF-center contrast-sensitive ganglion cells. The images on the left represent hypothetical responses of the cell to a spot of light presented to the retina in the center of the receptive field, in the surround of the receptive field, or in both the center and surround regions of the receptive field. A + symbol on the receptive field map indicates an increase in the firing rate of action potentials by the cell, that is, excitation; a – symbol indicates a decrease in the firing rate of the action potentials, that is, inhibition.

during illumination (ON cells) or no action potentials during illumination (inhibition) but a burst of action potentials at the cessation of illumination (OFF cells). With surround illumination the opposite response is elicited—an OFF response in the ON ganglion cells or an ON response in the OFF ganglion cells. When center and surround regions are illuminated simultaneously, the center and surround responses neutralize each other, and only a weak response is elicited in the cell that is reflective of the center response. In the central part of the primate retina, the ON- and OFF-center ganglion cells make up about 70% of the ganglion cells and in the periphery about 50% of the ganglion cells. Indeed, in the very central part of the primate retina, the fovea, the sustained ON- and OFF-center ganglion cells make up more than 90% of the ganglion cells.

Figure 4.9A is a scheme that suggests the basic circuitry underlying the formation of the ON- and OFF-center ganglion cells' receptive fields. The diagram describes only the simpler cone pathways. For example, many mammals do not have OFF-rod bipolar cells, and the mechanisms underlying surround antagonism in the rod pathway are not entirely clear. In the diagram, open circles indicate excitatory synapses, whereas filled circles show inhibitory-like junctions. In figure 4.9 the center responses of the ON- and OFF-bipolar cells reflect direct input from the center cone, which depolarizes the ON-bipolar cell and hyperpolarizes the OFF-cell. The inhibitory surround responses are mediated by the horizontal cells in the outer plexiform layer by feeding back onto the cones or forward onto the bipolar cells. Amacrine cells in the inner plexiform layer also contribute to surround responses by inhibiting the bipolar cell terminals and/or the ganglion cell dendrites. There is a difference between the horizontal and amacrine cell surround inhibition. Surround inhibition mediated by the amacrine cells has a finer grain than does horizontal cell–mediated surround inhibition, which is

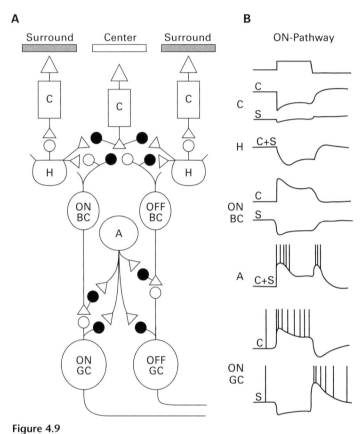

Figure 4.9
(A) A simplified scheme of the cone-mediated circuitry underlying ON-center, OFF-surround and OFF-center, ON-surround ganglion cells. Open circles indicate excitatory synapses; filled circles, inhibitory-like synapses. See text. (B) Idealized responses for ON-pathway cells when center (c) and surround (s) illumination is presented relative to the middle cone. See text.

more global. That is, surround inhibition mediated by the amacrine cells is more restricted in its extent.

Figure 4.9B shows idealized responses for cells in the ON pathway. Cones hyperpolarize when illuminated centrally (c) but show little response with surround illumination (s). Central illumination depolarizes the ON-bipolar cell, whereas both center and surround illumination (c+s) hyperpolarize the horizontal cell. On the other hand, surround illumination hyperpolarizes the ON-bipolar cells because of the inhibition mediated by the horizontal cells back onto the central cone or forward onto the bipolar cell.

The amacrine cell illustrated in figure 4.9 responds similarly to center and surround illumination, depolarizing at both the onset and offset of light. The ON-bipolar cell excites the ON-ganglion cell, but the amacrine cell can inhibit the ganglion cell response by inhibiting the bipolar cell terminals or the ganglion cell directly or both. The vertical lines superimposed on the amacrine and ganglion cell responses represent the action potentials generated by these cells. Note that the frequency of action potential generation depends on the extent of the underlying depolarization of the cell.

Blue/Yellow Color Opponent Ganglion Cells

As already noted, Old-World primates (from Africa), ourselves included, have three cone types—one that mediates long-wavelength (red) vision, a second that mediates medium-wavelength (green) sensitivity, and a third that mediates short-wavelength (blue) color vision. Color-opponent bipolar cells are found in many species; their receptive field center responds best to one color, and their surround responds to another color. In primates, however, this holds mainly for red/green bipolar cells. In other words, bipolar cells that get input from blue cones either have no surround response or a weak one. However, blue opponent ganglion cells that respond with excitation to blue light and that are inhibited by yellow light are found. Because there are no yellow-sensitive cones in the primate retina, the yellow sensitivity comes from a mix of the red and green cones. If a bipolar cell receives input from both red- and green-sensitive photoreceptors, it will be maximally sensitive to yellow light, and that is what happens. As shown by Dennis Dacey and Barry Lee, the blue/yellow opponent ganglion cell gets its blue input from ON-bipolar cells that receive their input exclusively from blue cones, whereas it receives inhibition from OFF-bipolar cells that receive input from both red and green cones. As expected, these blue/yellow ganglion cells are bistratified cells whose dendrites run in both the ON-layer, where they receive blue input, and the OFF-layer, where they receive input from both the red- and green-sensitive bipolar cells (figure 4.10).

Figures 4.9 and 4.10 suggest the synaptic circuitry underlying the generation of the receptive fields of two ganglion cell subtypes—the ON- and OFF-center and the blue/yellow opponent ganglion cells. As noted earlier, whereas the ON- and OFF-center cells are the most common ganglion cell types found in the primate retina, there are many other ganglion cell subtypes, perhaps more than a dozen. Currently much effort is being devoted

Figure 4.10
Circuitry underlying blue/yellow color opponency in the primate retina. The drawing illustrates a small bistratified ganglion cell that receives input in the ON-sublamina of the inner plexiform layer from a bipolar cell that receives input only from blue (B) cones. It receives input in the OFF-sublamina from diffuse OFF-bipolar cells that receive their input from green (G) and red (R) cones.

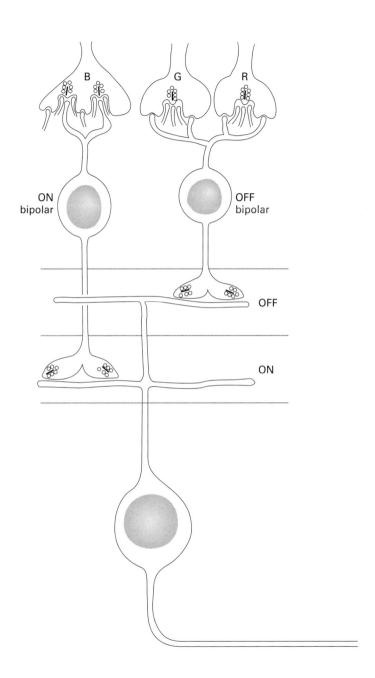

to analyzing these ganglion cell subtypes and the synaptic circuitry that underlies their responses.

Glaucoma

All visual information goes from the eye to higher visual centers via the retinal ganglion cells and their axons, which form the optic nerve. If the ganglion cells are lost or the optic nerve is disrupted, all information from the eye is lost. Such can happen in glaucoma, number 2 in the United States on the list of diseases causing blindness (after cataracts). Whereas light perception remains with those with severe cataracts, total blindness occurs in those suffering from severe, late-stage glaucoma.

Fortunately, early-stage glaucoma can be treated effectively. There are over 2 million people in the United States who have been diagnosed with glaucoma, and there are undoubtedly many more who have it but have not been diagnosed. Chronic glaucoma in its early stages does not cause any symptoms even though the eye is being slowly and steadily damaged with attendant loss of vision; thus, glaucoma is often referred to as the "thief in the night." The diagnosis of glaucoma can only be made as part of a complete eye examination. A major reason the early loss of vision is not recognized is that the initial visual loss is usually in the visual field periphery, the blind area in one eye is seen by the corresponding area of the other eye, and so early changes in peripheral vision are not noted. And, of course, peripheral vision is not sharp as is central vision. Glaucoma destroys central vision last, so patients with advanced glaucoma can often see small objects they are looking at directly. As noted in chapter 1, it is possible in some instances to have 20/20 vision and be legally blind because of extensive loss of peripheral vision.

Figure 4.11 shows visual fields of a glaucoma patient with a normal field in the right eye (A) and the same patient with a large scotoma (area of blindness) in the inferior retina of the left eye (B). The areas where the patient does not see the test lights are black or gray. In the normal eye (A) the black/gray area on the right is where the optic nerve leaves the eye, the blind spot, where there are no photoreceptors. In the right eye virtually the entire inferior right visual field is blind, as well as an area in the left inferior retina along with the blind spot.

Glaucoma is a disease manifested by damage and ultimately destruction of the ganglion cells and optic nerve. A very strong risk factor is elevated intraocular pressure (IOP). Other risk factors include heredity (a person with a family history of glaucoma is considered a lifelong glaucoma suspect), advancing age, Asian, African, or Hispanic ethnicity, and hyperopia. The diagnosis is usually made by the presence of elevated IOP and/or evidence of optic nerve or ganglion cell damage. To make the initial diagnosis difficult, some eyes resist damage in the presence of mildly above-normal IOP (ocular hypertension), whereas other eyes sustain glaucomatous damage in the presence of normal IOP (normal tension- or low-tension glaucoma). Some experts suggest that eyes with normal-tension glaucoma have elevated IOP during times such as the middle of the night when testing is not ordinarily done or that even normal IOP is damaging to the ganglion cells. Recent measurements of IOP throughout the day and night showed there are brief, spike-like increases in IOP of substantial magnitude, and these might be responsible for damaging ganglion cells in normal-tension glaucoma.

There are two basic types of glaucoma. The primary glaucomas are congenital (present at birth and often associated with enlargement of the entire eye), chronic (primary open angle), and acute (narrow angle). Secondary glaucomas are caused by

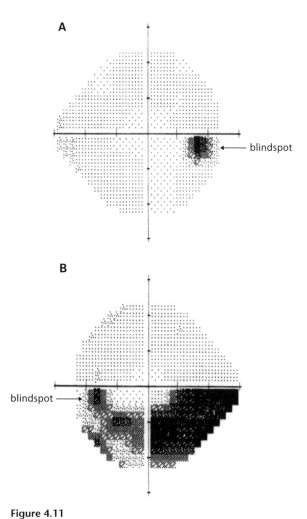

Figure 4.11
Visual fields of a patient with a normal field in the right eye (A) (the dark
spot is the normal blind spot where the optic nerve leaves the eye) and
the left eye (B) with an extensive inferior glaucomatous scotoma (black
area).

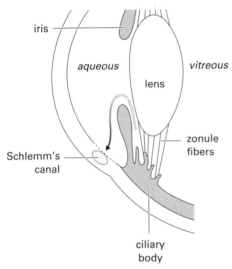

Figure 4.12
Circulation of the aqueous humor. Aqueous humor is produced by the ciliary processes in the posterior chamber and drains from the anterior chamber via the canal of Schlemm.

other eye conditions such as intraocular inflammation, certain drugs, trauma, and intraocular hemorrhage.

The normal human IOP is relatively high, 15–20 mm of mercury; just press on your eye to judge the firmness. The pressure is high enough to keep the eye from collapsing and to provide a firm structure for incoming light rays to be sharply focused. The IOP is maintained by the aqueous humor (see figure 4.12). The aqueous is secreted by the ciliary body in the posterior chamber (the space between the back of the iris and the front of the lens) and circulates through the pupil into the anterior chamber and exits the eye through tiny drainage channels (canals of Schlemm) located in the periphery of the anterior chamber. Open-angle glaucoma occurs when the outflow channels are unable to drain the aqueous inflow adequately. Formation of aqueous continues, and the IOP rises. Whereas primary open-angle glaucoma causes no initial sign or symptoms and develops gradually over long periods of time (months or years), acute glaucoma comes on suddenly, often in a matter of minutes, and causes dramatic symptoms including severe eye pain, striking loss of vision, intense eye redness, marked head pain, and other systemic symptoms. The IOP may suddenly rise from a normal of 15–20 to 50, 60, or even 100 mm of mercury. Acute glaucoma occurs in eyes when access to the outflow apparatus is suddenly blocked, by the iris when, for example, the pupil dilates in darkness or for other reasons such as extreme stress or certain medications.

If intraocular pressure rises quickly and significantly, as in acute glaucoma, intraretinal blood circulation can be compromised, and neurons, especially the ganglion cells in the inner retina, are damaged and can die. The outer retina is primarily served by the choroidal circulation situated behind the retinal pigment epithelium and so is usually spared. But why the ganglion cells and their axons are especially susceptible to modest increases in IOP that do not greatly affect retinal circulation, as in primary open-angle glaucoma, is not well understood. One idea is the following: ganglion cell axons run along the surface of the retina unmyelinated—that is without the glial cell sheath that is typically wrapped around nerve cell axons, insulating them and promoting faster conduction speed. The axons gather at the back of the eye to form the optic nerve. The optic nerve axons, on leaving the eye, must pass through openings in a structure called the cribriform plate, and as they emerge from the plate, they then acquire myelin. On the surface of the cribriform plate, and even extending into the retina and along its inner surface, is a type of glial cell called an astrocyte (myelin is formed by a glial cell called an oligodendrocyte, not found in the retina or anterior to the cribriform plate). There is some evidence that increased intraocular pressure can activate astrocytes, and whereas reactive astrocytes can be beneficial in certain situations, they also can be harmful. One theory is that the reactive astrocytes clog the openings in the cribriform plate, compressing and damaging ganglion cell axons as they pass through the plate. When their axons are severely damaged, ganglion cells die. There is also a possibility that reactive astrocytes release substances that can cause the death of ganglion cells.

All ganglion subtypes may not be equally susceptible to glaucoma. It appears that the larger ganglion cells that are particularly responsive to movement are the first cells to show damage. Why this is so is not clear, but this fact may have a future use in diagnosing glaucoma in its very early stages. Tests that examine the responses of small areas of the retina to moving or adapting stimuli are being developed, and these tests show promise for an earlier diagnosis of the disease.

The diagnosis of glaucoma is made by measurement of IOP (tonometry) and a visual inspection of the optic nerves and ganglion cells for damage. The latter is accomplished by three-dimensional examination of the optic nerve, by measurement of the thickness of the ganglion cell layer, and by charting the peripheral visual fields of vision for areas of vision loss (see figure 4.11).

As noted, glaucoma can be effectively treated especially in its early stages. Clinical studies have clearly demonstrated that lowering IOP will usually prevent eye damage and visual loss even in low-tension glaucoma. The traditional first line of treatment is the daily use of pressure-lowering drops, which are usually effective. The second line, if drops alone do not suffice, is laser treatment of the outflow channels, and finally, in the unusual cases where these procedures do not adequately control IOP, surgery is required.

The treatment of acute glaucoma (a true ocular emergency), in which the outflow area of the anterior chamber is suddenly blocked by the iris, is the instillation of iris constriction drops and a surgical (laser) procedure to place a hole in the peripheral iris to equalize the pressure between the anterior and posterior chambers. Patients with chronic glaucoma ordinarily must continue drop treatments for life, whereas patients with acute glaucoma may be cured by prompt surgical treatment. Congenital glaucoma is essentially a surgical disease that, fortunately, is usually successful. The important take-home message regarding glaucoma is that treatment is preventative, so early diagnosis and treatments are critical. Vision lost from glaucoma cannot be restored. A patient blind from cataracts can have vision restored to normal by surgery, but a patient blind from glaucoma is permanently blind.

Diabetic Retinopathy

A total of 22.3 million Americans (7% of the population) are currently diagnosed with diabetes in the United States at an estimated cost of $245 billion in 2012 alone, and a surge in the number of people with diabetes is expected in the next 20 to 30 years propelled by the current epidemic of obesity. The surge in diabetes is a worldwide phenomenon. A recent Chinese study of about 100,000 adults reported an incidence of 12% for diabetes and 50% for prediabetes (those with above-normal blood glucose levels). A serious complication of diabetes is damage to the small blood vessels of the retina, called diabetic retinopathy. Diabetic retinopathy is not a disease of a specific type of retinal cell, as is glaucoma. Rather, present views suggests that it results from abnormally high blood sugar coupled with the high energy demand of the retina, especially that of the photoreceptors. The severity of diabetic retinopathy ranges from no effect on vision to complete blindness. The National Eye Institute has reported that nearly 50% of all patients with diabetes have at least some degree of diabetic retinopathy.

There are two types of diabetes: type 1 usually develops in childhood and is more severe than type 2, which is usually adult in onset but is being seen more frequently in young people. Virtually all type 1 patients develop diabetic retinopathy in about 7 years on average after the onset of diabetes. It gradually but slowly progresses to severe diabetic retinopathy. The time from onset to retinopathy in type 2 diabetics is less settled, and about 30% of type 2 diabetics do not develop retinopathy. Some type 2 patients (usually those who do not have periodic physicals with blood sugar tests) have retinopathy when the diagnosis of diabetes is first made, meaning the diabetes had been present but undiagnosed for a considerable time. On some occasions the diagnosis of diabetes is made during an eye examination.

The earliest observable stage of diabetic retinopathy, known as nonproliferative retinopathy, is characterized by localized dilations in weakened retinal blood vessels called microaneurysms (figure 4.13). Dilated retinal vessels and small dot hemorrhages are also often seen, but there is even evidence of abnormal neural function before the vascular changes are seen. The first significant effect on vision occurs when fluid accumulates in the macula, causing swelling known as macular edema.

Advanced diabetic retinopathy, known as proliferative retinopathy, is characterized by the growth of new blood vessels, usually into the vitreous humor. These new blood vessels are fragile and tend to bleed, often causing sudden and severe visual loss, or they may adhere to the retina, pulling it off the back of the eye when the vessels contract, causing retinal detachment.

What leads to diabetic retinopathy is not well understood, but it is known that the elevation of blood sugar (glucose) levels is critical for the initiation of the disease and that tight control

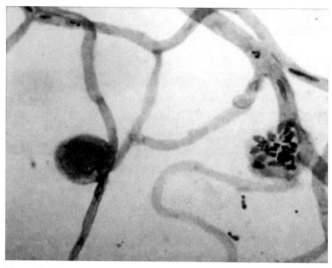

Figure 4.13
Retinal capillaries from a patient showing the early changes in diabetic retinopathy. Two microaneurysms are obvious; one contains observable blood cells (right).

of blood glucose levels significantly inhibits its development and progression. Why increased levels of glucose lead to blood vessel and neuronal alterations is not understood, but one theory is that elevated blood glucose induces hypoxia (decreased oxygen) in the retina.

Several findings relate hypoxia to the initiation of diabetic retinopathy. As noted above, before there are any observable vascular changes in patients, there are signs of abnormal retinal function. Dark adaptation is slow, and there is a loss of color discrimination as well as other visual defects, all of which can be reversed, at least partially, by the inhalation of oxygen. Further, diabetic patients who have lost substantial numbers of photoreceptors because of retinal degenerative diseases such as

A

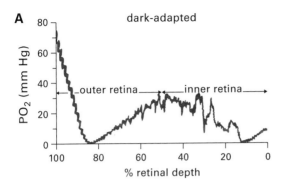

B

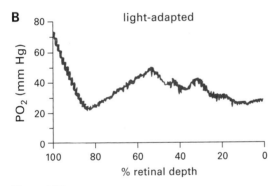

Figure 4.14
Profiles of PO_2 as a function of percentage retinal depth from the central retina of a monkey in dark adaptation (A) and light adaptation (B). The minimal PO_2 at ~80% retinal depth corresponds to the inner segment region of the photoreceptors, where the Na^+ pumps are located.

retinitis pigmentosa do not develop diabetic retinopathy, presumably because the energy demand of the retina is reduced by the photoreceptor loss. Recall that photoreceptors are most active in the dark, and light turns them off. A single mouse rod, for example, consumes 10^8 ATP molecules per second when fully dark adapted. Figure 4.14 shows levels of O_2 measured in the central region of a macaque monkey first dark adapted and then exposed to light. Whereas the levels of O_2 are high at the tips of photoreceptor outer segments, they rapidly plunge to virtually zero near the inner segment region of the retina. The reason for this is that the inner segments contain the energy-demanding pumps that rid the photoreceptors of the ions that flood into the cell in the dark. Levels of O_2 are higher in the inner retina but again show a dip in the inner plexiform layer, presumably as a result of sustained activity especially of the OFF-cells in the dark. After light exposure the levels of O_2 are higher throughout the retina as the photoreceptors are shut down by light, as well, presumably as a result of neurons in the inner plexiform layer shutting down, that is, OFF-cells.

In support of the idea that hypoxia relates to diabetic retinopathy and the early neuronal changes that take place, Arden and his co-workers have recently shown that by partially suppressing rod activity with dim light at night in humans, some amelioration of diabetic retinopathy occurs. Another observation is that extensive laser photocoagulation, which destroys both photoreceptors and retinal neurons, is protective against diabetic retinopathy. Such photocoagulation is commonly used as a treatment for severe diabetic retinopathy; by reducing the metabolic activity of the retina and its oxygen demand, progression of the disease can be slowed or even stopped.

How might high levels of blood glucose and hypoxia be related? Again the story is incomplete, but there is some evidence that in the presence of elevated glucose, retinal glial cells, especially the Müller cells, produce substances that prevent retinal blood vessels from dilating. Ordinarily, blood vessels throughout the brain dilate when there is an increased need for oxygen, but in diabetes this does not happen in the retina, and this could contribute to hypoxia. With hypoxia, as noted above, retinal tissue is compromised, leading to alterations in both the retinal blood vessels and the neurons and to the initiation of full-blown diabetic retinopathy as described above.

Another idea as to an underlying cause of diabetic retinopathy comes from recent experiments from Margaret Veruki's laboratory in Norway that have shown in a rat model of diabetes that there are changes in the release of the inhibitory neurotransmitter GABA in the inner plexiform layer from one subtype of amacrine cell that interacts with the terminals of rod bipolar cells. The authors propose that this perturbation of the rod circuitry could explain the decrease in rod sensitivity observed early in diabetic retinopathy and the slowing of dark adaptation. How the increased blood glucose level causes these effects is not clear at the present time, nor whether hypoxia is at all involved, but these experiments point the way to our finding this out.

Diabetic retinopathy is a treatable condition. The first principle of treatment is tight control of blood glucose by careful attention to diet, weight control, and by the use of insulin and other glucose-controlling medications. Also important is control of ancillary conditions such as high blood pressure and blood lipids. Early laser treatment can also prevent the advance of retinopathy and help maintain good visual acuity. Treatments for advanced diabetic retinopathy include ocular injections of anti-VEGF agents and steroids, and vitrectomy, the surgical removal of the vitreous humor. The latter is indicated when there is an extensive vitreous hemorrhage and evidence of extensive proliferative retinopathy. Bariatric (weight loss) surgery has proven to be effective in controlling obese patients who have difficulty maintaining tight glucose control.

5   Beyond the Retina—Lateral Geniculate Nucleus and Visual Cortex:

Amblyopia

The occipital cortex, the destination of much of the retinal output and the seat of visual perception, is considerably more complex than the retina and has been difficult to analyze in terms of synaptic circuitry. However, we do know something about the analysis of visual information beyond the retina in mammals, especially in the cat and monkey. The activity of individual neurons throughout much of the mammalian cortex has been recorded in these animals, but these recordings have been, for the most part, extracellular recordings that pick up the large action potentials generated by the neurons (and virtually all cortical neurons generate action potentials) but not the underlying depolarizing and hyperpolarizing potentials that reflect the synaptic input to the cells. The extracellular recordings have told us much about the coding of visual information by neurons, and this, of course, is reflected in the receptive field properties of the cortical neurons.

Lateral Geniculate Nucleus

In cat, monkey, and humans, most retinal ganglion cells project to the lateral geniculate nucleus (LGN) in the thalamus. LGN neurons then carry the visual message to area V1 of the cortex (figure 1.2). Why does this happen? Why do retinal ganglion cells not project directly to the cortex? The LGN, in addition to receiving retinal input, also receives substantial input back from the cortical areas concerned with vision as well as from a region of the brain involved in regulating activity levels throughout the brain. This brain region, called the reticular formation, plays an important role in mediating arousal states of the brain and in controlling levels of attention. Thus, the LGN serves to modify or regulate visual information before it is passed on to the visual cortex.

An example of how attention can affect the responses of a cortical neuron is shown in figure 5.1, an experiment carried out by Robert Wurtz at the National Eye Institute. Wurtz observed that when a monkey focuses its attention on an object in the receptive field of a peripheral neuron being recorded, the responses of the cortical neuron are significantly increased. The monkey is trained to look continuously at a fixation point straight ahead, but then a light comes on to the side which is in the receptive field of the neuron being recorded. Initially, the response of the neuron is modest. If the monkey is trained to attend to the peripheral stimulus, but not to look directly at it, and then to reach out and touch it, the activity of the cortical neuron is greatly increased. This effect appears to be mediated in large part by the LGN.

Figure 5.1
The effect of attention on the responsiveness of a cortical neuron. (A) The neuron being recorded is activated by a visual stimulus in the periphery of the visual field. When the monkey is looking at a fixation point in the center of its visual field, the response of the neuron to the peripheral stimulus is minimal. (B) If the monkey attends to the peripheral stimulus by touching it, the response of the neuron is significantly enhanced even though the monkey has not shifted its gaze from the fixation point.

A

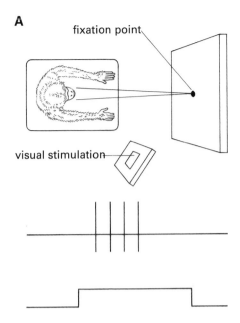

fixation point

visual stimulation

B

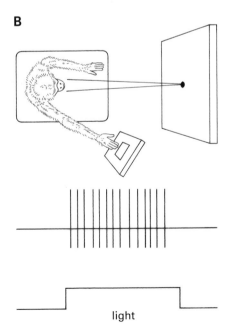

light

The receptive field properties of the LGN neurons are quite similar to those of the ganglion cells. The great majority of the retinal ganglion cells projecting to the LGN are the ON-center, OFF-surround and OFF-center, ON-surround types. These are the most common types of ganglion cells found in the cat and monkey retinas, and they come in two subtypes. Most show more-or-less sustained center and surround responses (figure 4.8), and these cells have small receptive fields. A minority of ganglion cells show transient ON- and OFF-center and surround responses and are quite movement sensitive (although not directionally sensitive); these cells have large receptive fields. Figure 5.2 shows receptive field maps for retinal ganglion cells and LGN neurons.

The LGN is a highly layered structure, especially in primates. Four dorsal layers receive input from the small, more sustained ON- or OFF-center cells, whereas two ventral layers receive input from the larger, more transient, movement sensitive ON- or OFF-center ganglion cells. There are also some cells between the layers, and they tend to be the blue/yellow opponent and other relatively rare ganglion cells. This is the beginning of a segregation of cells responding to color, movement, and form in the LGN—something that is particularly prominent in the cortex.

Eye Movements and the Superior Colliculus

Not all retinal ganglion cells project to the LGN, but some go to subcortical structures where they provide information to allow those areas to perform tasks important to vision and other allied phenomena. One such area is the superior colliculus, which is involved in the control of rapid eye movements essential for altering eye position so that images are brought to bear on the fovea for detailed viewing. Indeed, when we inspect an image, our eyes dart around the image, stopping briefly to inspect those aspects of the image of interest, as first shown by Alfred Yarbus

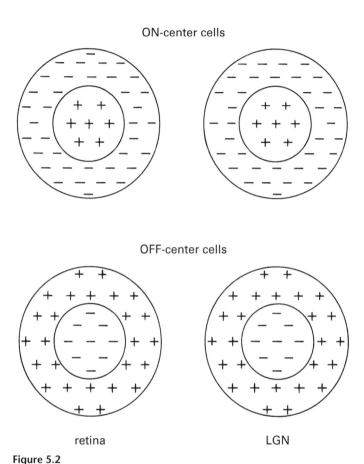

Figure 5.2
Receptive field maps of ganglion cells in the retina and LGN. Both ON and OFF ganglion cells and LGN neurons have a center-surround receptive field organization. A + sign means that the cell generated action potentials when the light went on in that location; a – sign means that the cell was inhibited from generating action potentials when a light went on in that spot but did generate action potentials when the light went off.

Figure 5.3
Eye movements of an observer scanning a face. The eyes jump around and stop momentarily on areas of interest, especially the eyes. These pictures were originally published by A. Yarbus.

in the 1950s. Figure 5.3 shows how our eyes view a face and how they stop to look carefully at the eyes, mouth, and other features.

These rapid eye movements are called saccades, and humans typically make two to three saccadic eye movements per second. Between saccades the eye maintains fixation on whatever is focused on the fovea for a quarter to half a second, which is sufficient time to appreciate the part of the image in view. During a saccade, the eye moves rapidly, but we are not aware of the movement. Indeed, there is good evidence that vision is suppressed during a saccade.

In addition to input from the retina, the superior colliculus receives abundant input from the cortex. A simplified way to think about how we view an image is as follows. The cortex decides what we want to look at next, and the superior colliculus commands the appropriate saccade by activating neurons that control the eye muscles, of which there are six. Four of the muscles (two antagonistic pairs) move the eyes primarily in the horizontal or vertical directions, whereas the last two muscles are involved primarily in rotating the eye. The ensuing saccade moves the eye such that it brings what the cortex wants to view onto the fovea.

In addition to saccades, there are two other types of eye movements not controlled by the superior colliculus that are important. Smooth pursuit eye movements allow us to track moving objects or to focus on a fixed object when we are moving. Vergence eye movements allow the eyes to focus on near and far objects—to see a near object with both eyes requires the eyes to

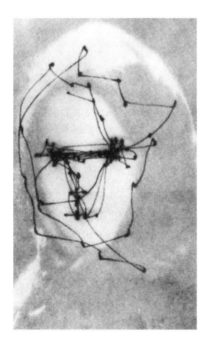

converge, whereas to see a distant object, the eyes must diverge to become parallel. Several areas and pathways are involved in mediating these two types of eye movements.

A final point with regard to eye movements is that the eye is always moving somewhat, and if an image is stabilized on the retina, the image fades away. This was first shown by Troxler in 1804, and more recent experiments in the 1950s by Loren Riggs and Floyd Ratliff as well as others clearly confirmed Troxler's observations. These movements can be small saccades, slow drifts, or even a tremor of the eye.

Returning to the superior colliculus, it is of interest to note that in nonmammalian species this brain area is called the optic tectum, which is responsible for much of the sensory-motor integration in these animals and for initiating their behavioral responses. As the cortex expanded during evolution, it took over most sensory-motor integration and the initiation of motor movements, leaving the superior colliculus to mediate primarily saccadic eye movements.

Visual Cortex—Area V1

LGN neurons project to area V1 in the posterior cortex (figure 1.2). In V1 there is clear evidence of further processing of visual information. A hierarchy of receptive field responses is observed in V1 that relate to the stimuli that best activate the neurons. In the case of cortical neurons it is not simply the intensity of light shone on the retina that best activates a cell but the shape and orientation of the stimulus projected on the retina as well as whether it is moving and in what direction. To an optimal stimulus a cortical neuron responds vigorously with a large burst of action potentials. To a suboptimal stimulus the cell responds weakly or not at all.

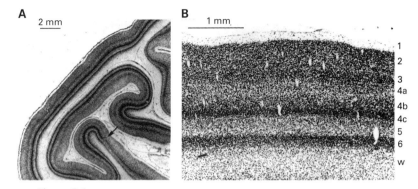

Figure 5.4

Low- (A) and high-magnification (B) light micrographs of sections through area V1 of the visual cortex. The cortex is about 2 mm thick and is layered (A). The six major layers that are distinguished consist of cells of different packing densities (B). Below layer 6 is white matter (W), which contains myelinated axons that carry information to and from specific cortical regions and out of the cortex. The arrow in A indicates the border of visual areas 1 and 2.

Area V1, like the retina and LGN, is a layered structure (figure 5.4A). Six layers are recognized based mainly on cell density packing (figure 5.4B). The thickest layer is layer 4, and it is subdivided into three sublayers, a, b, and c. The input from the LGN mainly comes into layer 4, and where in layer 4 depends on its origin. Input from the four distal LGN layers goes to layers 4a and 4c (its lower part), whereas the input from the proximal (movement-sensitive) two layers goes to layer 4c (its upper part). Interestingly, the blue/yellow color cells bypass layer 4 and project directly to layers 2 and 3. The cells in layer 4 project to the other cortical layers (except for layer 1, which has few cells; Figure 5.4B).

In primates there are a few cortical cells in layer 4 that have a center-surround receptive field organization like retinal and

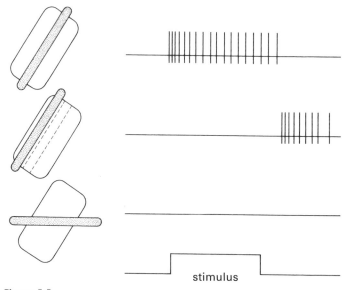

Figure 5.5
Responses of a simple cell to a bar of light. When the bar covers the excitatory zone of the receptive field, the cell fires vigorously (top). When the bar stimulates the inhibitory zone, a strong off-response is recorded (middle). When the orientation of the bar is dissimilar to that of the receptive field, the cell fires very weakly or not at all (bottom).

LGN neurons, but most of the cells recorded in and around layer 4 have a more elaborate receptive field organization, as first shown by David Hubel and Torsten Wiesel in the late 1950s. Whereas stationary spots of light projected onto the retina will activate the cortical cells weakly, a much more effective stimulus is a bar of light that has a specific orientation. These orientation-specific cells, called simple cells, have elongated receptive fields that, however, can be mapped into excitatory and inhibitory regions (figure 5.5). A bar of light falling on the excitatory region strongly excites the cell, whereas the bar falling on the

A   B

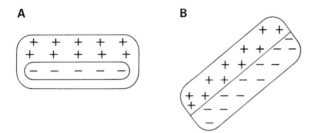

Figure 5.6
Receptive field maps of other simple cortical cells. In both cases an oriented bar or edge is the optimal stimulus. (A) The central zone (inhibitory in this case) is eccentrically positioned. (B) The excitatory zone is on one side of the field, and the inhibitory zone on the other.

inhibitory region strongly inhibits the cell. If the orientation of the bar is altered from optimal, both the excitatory and inhibitory regions are stimulated, and the cell usually fails to generate any action potentials at all.

How much variation in the orientation of a bar is permitted? A change in orientation of about 10° can be detected by a typical simple cell, so to go from one orientation to the same orientation is a change in orientation of 180°. This means that for any bit of the visual field there are 18–20 or so neurons receiving input that differ with regard to optimal orientation selectivity.

There are also simple cells whose receptive fields differ somewhat from the one illustrated in figure 5.5. In these cells the central excitatory or inhibitory zone is eccentrically placed in the receptive field, or the excitatory zone is on one side of the field and the inhibitory zone on the other (figure 5.6). The latter cell type obviously responds best to a light-dark edge of light. Again, these cell types have various orientation requirements, and orientations that differ from the optimal orientation by more than 10° decrease or completely inhibit the response of the cell.

Complex Cells

Simple cells are generally recorded close to layer 4, the input layer. Further away from layer 4, in layers 2, 3, 5, and 6, are cells that cannot be activated at all with small spots of light falling on the retina. To activate these cells, termed complex cells, requires an oriented bar of light moving across the cell's receptive field. Further, the bar must be moving at right angles to the orientation of the bar.

Some of these complex cells, especially in layer 2 (10–20%), are directionally selective; a bar moving in one direction strongly excites the cell, whereas the bar moving in the opposite direction inhibits the cell. Other specialized complex cells require bars of light of specific length moving through the receptive field; such cells are called end-stopped cells.

An obvious conclusion that can be drawn from the above results is that, as one moves along the visual pathway, the stimulus presented to the retina must be more and more specific to drive the cells optimally. In the retina and LGN, spots of light appropriately positioned drive the cells well, but neurons around layer 4 in the cortex require an oriented stimulus—a bar or edge of light—to maximally stimulate them. Further away from layer 4, an oriented bar or edge is required that is moving in a particular direction. And even further away, the moving stimulus bar must be moving in only one direction and/or be limited in its length. Put in other terms, as one moves along the cortical visual pathways, specific aspects of the visual image are encoded in individual cells—first orientation, next orientation and movement, then movement in a specific direction, and so forth.

Formation of Cortical Receptive Fields

Although the synaptic circuitry of the cortex is not yet worked out, one can speculate as to how cortical neuron receptive fields might be structured from the LGN input neurons or from synaptic interactions among the cortical neurons themselves. For example, if ON-center LGN neurons receive input from ganglion cells that are aligned in a particular orientation on the retina, and they all feed into one cortical cell as shown in figure 5.7, that cortical cell would have the properties of a simple cell. The overall receptive field size of simple cells is much larger than the receptive field size of an LGN neuron, which provides support for this idea.

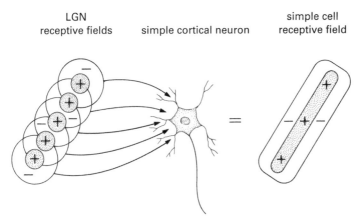

LGN receptive fields   simple cortical neuron   simple cell receptive field

Figure 5.7
How a simple cortical cell's receptive field could be fashioned by excitatory inputs from the LGN. The receptive fields of the input neurons (ON-center cells) overlap and lie along a straight line on the retina (left). The receptive field of a cortical neuron receiving such input will consist of an elongated central excitatory zone surrounded by an inhibitory zone (right).

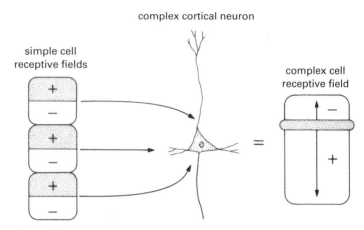

Figure 5.8
How a directionally selective complex receptive field could be formed from edge-selective simple cells. If the receptive fields of the input neurons are aligned on the retina as shown on the left, downward movement of a bar will elicit vigorous activity in the complex cell, whereas upward movement will result in no response. See text.

Directionally selective complex cells could be generated from edge-selective simple cells that provide input to a complex cell. If the edge-selective simple cells are aligned as shown in figure 5.8, a bar moving in the downward direction would first encounter the excitatory zones of the simple cells' receptive fields, and the complex cell would be strongly stimulated. A bar moving in the opposite, upward direction would activate the inhibitory zones first, thereby inhibiting the excitatory zone, and no response would occur in the directionally selective complex cell.

The schemes shown in figures 5.7 and 5.8 are undoubtedly overly simplified, but they do suggest a way to think about possible cortical synaptic organization and how the hierarchy of cortical cell types may be formed.

Binocular Interactions

The advantage of having two eyes whose visual fields overlap is the capability of depth perception. This is easily demonstrated by trying to oppose index fingers if one eye is closed and your head is still; more often than not, you will fail this task. To accomplish this task easily and reliably, input from both eyes to the same cortical neurons is required, and most neurons in V1 receive input from both eyes—they are binocular. Indeed, it is in V1 that binocular cells are first encountered. Cells in the LGN are monocular, receiving input from one or the other eye. (Each of the layers in the LGN receives input from only one eye.)

It has been demonstrated, however, that cortical cells are most often driven more strongly by one eye than the other, a phenomenon termed ocular dominance. This is shown in figure 5.9 in an experiment carried out by Hubel and Wiesel in the cat. It shows the relative number of cells encountered when recording cells throughout the cat cortex as a function of whether the cell responds to only one eye (bins 1 and 7), equally from both eyes (bin 4), or to intermediate levels of input from one eye or the other (bins 2 and 3 and bins 5 and 6). The cells receiving input from just one eye (monocular cells) are found in layer 4, the input layer of the cortex; the most binocular cells (bins 3, 4, and 5) are found more distally—in layers 2 and 6. Thus, there appears to be a hierarchy of degree of ocular dominance just as there is in terms of receptive field complexity.

When a cell receives input from both eyes, both eyes must be stimulated in the same way and in the corresponding retinal location. If an oriented bar of light moving in one direction is required to best activate a cell receiving input from one eye, the exact same stimulus is required to drive the corresponding cell in the other eye. If both eyes are stimulated together with the same stimulus, the response of the cortical cell is greater than if just

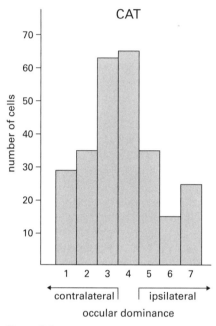

Figure 5.9
Ocular dominance histograms for cortical cells in the cat. Cells in groups 1 and 7 are monocular (driven by one eye), whereas cells in group 4 are equally binocular (driven as well by both eyes). Cells in groups 2, 3, 5, and 6 are binocular but driven more by one eye or the other.

one eye is stimulated. However, the response of the cell is usually greater from one eye than the other, showing ocular dominance, (unless the cortical cell is driven equally by both cells).

How does binocularity of the cortical cells relate to depth perception? Although all binocular cortical cells have their receptive fields in comparable retinal areas in the two eyes, most respond well even if the stimuli presented to the two eyes are not perfectly aligned. Certain complex cells, however, do require precisely aligned stimuli to be presented to the two eyes for the cells to respond. Termed disparity-tuned cells, these cells, first described by Horace Barlow, Colin Blakemore, and Jack Pettigrew in the late 1960s, are believed to be critical for depth perception.

Because our eyes are separated, images near or far fall on corresponding but slightly different parts of the retina. For most cells in V1, this makes little difference. For disparity-tuned cells, this difference is critical. Some respond only when images are near, others only when images are further away. In other words the disparity-tuned cells require exactly positioned images on the retina for them to respond optimally.

Cortical Organization—The Hypercolumn

There is much going on in each part of area V1. There are cells with different receptive field properties and that appear hierarchical in nature. Simple cells come before complex cells, which come before more specialized complex cells. But all of the V1 neuronal receptive fields described so far (except for a few in layer 4) have a requirement for orientation. There are V1 cells that receive input from only one eye; however, most receive input from both eyes but to varying degrees. And there are V1 cells interested primarily in color, which are to be discussed.

Hubel and Weisel proposed that there is a basic cortical organization in area V1 that takes into account all of this complexity. That is, the cortex is organized into modules, called hypercolumns, that have the dimensions of 1 mm by 1 mm by 2 mm, the latter being the thickness of the cortex. This basic cortical structure contains all the neuronal machinery necessary to analyze a bit of visual space. These are not completely separate modular units, but they overlap.

The organization of a hypercolumn is shown in figure 5.10. First, input from the two eyes comes into layer 4 of the cortex separately, on either side of the hypercolumn, forming columns that run across the cortex in layer 4, forming somewhat irregular stripes. The columns are ~0.5 mm wide, so that to encompass information from both eyes requires 1 mm—one dimension of the hypercolumn. Above and below where the LGN input comes in, the cells are binocular, the extent of which depends on how far the cell is from layer 4. Cells in layer 4 usually show no binocularity; they are monocular as indicated by a circle on the left side of figure 5.10. The degree of binocularity is indicated by + symbols, and three + symbols indicate the cell is receiving roughly equal input from both eyes.

A similar organization is seen for cells with different receptive field properties (right side of figure 5.10). A few cells in layer 4 may have a center-surround receptive field organization (CS), whereas simple cells are found above and below these cells. Complex cells (C) are found further away from the LGN input, and specialized complex cells (SC) further still.

Running roughly at right angles to the ocular dominance columns are orientation columns. Because there are 18–20 or so different orientations that cells can prefer, these columns are much narrower, about 0.05 mm in width. To take into account all the orientation possibilities requires about 1 mm of the cortex, the second dimension of the hypercolumn. The orientation columns are clearest at the edges of the ocular dominance columns. The reason for this is that in the centers of the ocular dominance columns are groups of cells that have color preference, and the orientation columns converge into the color areas as shown in figure 5.10. The grouping of color cells does not go all the way through the thickness of the cortex; rather, they are found above and below layer 4 and often are referred to as pegs or blobs. The cells in the color pegs have a center-surround receptive field organization and may be either single opponent or double opponent (see chapter 4). They are the only cortical cells outside of layer 4 not to have an orientation preference.

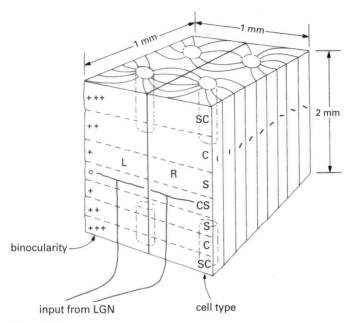

Figure 5.10

A hypercolumn: a 1 mm × 1 mm × 2 mm block of cortex containing all the cells required to analyze a bit of visual space. In a hypercolumn, input from both eyes is represented. Furthermore, all types of simple and complex cells are present, and the cells have all possible orientation preferences and varying degrees of binocularity. In addition, color-sensitive cells are found in the pegs inserted into the hypercolumn. Degree of binocularity is represented by (+) symbols; in layer 4, where lateral geniculate input enters the cortex, the cells are monocular (represented by an open circle); away from layer 4 the cells are more and more binocular. Cell types also vary through the thickness of the cortex. Around layer 4, some cells are center-surround (CS); away from layer 4c they are first simple (S), then complex (C), and finally specialized complex (SC). See text.

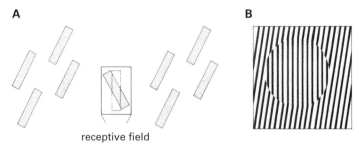

A   B

receptive field

Figure 5.11

(A) The effects of distal stimuli on the response of a cortical neuron. The orientation tuning of a cell can be altered by up to 10° by the presence of oriented stimuli outside the receptive field. The orientation selectivity of the neuron (clear bar) is tilted to the left (shaded bar) when bars of light tilted to the right are presented outside the receptive field. (B) The tilt illusion. The lines in the circled area appear tilted to the left although they are vertical.

The cortex, then, consists of a series of repeating modules, each looking after a bit of the visual field. Whereas the major organizing principle of the cortex is vertical, there are also horizontal connections between the modules that link cells within a specific layer. This means, for example, that information is integrated over many millimeters of cortex, and thus, a cell in one module can be influenced by stimuli falling on other modules. A revealing example is shown in figure 5.11, where the orientation of lines appears altered by the orientation of surrounding lines. Known as the tilt illusion, the lines in the center appear tilted to the left although they are perfectly vertical.

Amblyopia

A common condition that involves the visual cortex is called amblyopia, which is the most frequent cause of poor vision in childhood, present in 2–4% of children. Amblyopia is defective vision in an eye that is structurally normal. The visual acuity of newborn infants has been measured as 20/400, less than legal blindness. The main reason for this is that the fovea is not well developed at birth; indeed, it takes several months for the fovea to develop fully. But it is also the case that sharp visual images must fall on the photoreceptors early in life for clear high-acuity vision to develop. If this does not happen, for whatever reason, the development of clear foveal vision is severely compromised, the condition of amblyopia. The defect, however, is not in the eye but in the cortex. There is, therefore, a critical period for cortical maturation that extends from a few months after birth to about 6–8 years of age. Any significant interference with, or blurring of, the visual image up to this time can result in amblyopia. The common causes of amblyopia are strabismus (i.e., crossed eyes), opacities in the optic media (i.e., congenital cataract), or substantial refractive error (i.e., high astigmatism, high myopia, or hyperopia) in one eye.

Strabismus is an eye muscle imbalance in which the two eyes are not aligned. One eye can be turned in (esotropia), out (exotropia), up (hypertropia), or down (hypotropia). Babies and young children with strabismus normally have both eyes open and appear to be looking with both eyes. Actually they are viewing an image with only one, the straighter eye. Babies and young children have the inherent ability to suppress the image from the nonstraight eye. Otherwise the brain would receive two distinct disparate images, and objects being observed would appear double (diplopia). The capability of suppression eliminates diplopia. On the other hand, adults do not have this suppression capability. When they develop strabismus, which is a not uncommon occurrence as a result of disease such as diabetes, myasthenia gravis (a neuromuscular disease), orbital injury, or vascular occlusion, they have persistent diplopia. This condition can be intolerable and only be relieved by occluding one eye with a patch, special glasses, or eye muscle surgery. Although most patients with strabismus have one straight dominant eye, in some patients the fixation of the eye alternates, (particularly those with exotropia). When looking to the right, they use one eye, to the left the other eye, and acuity is equivalent in both eyes. They have the ability to suppress the visual input from the eye they are not using at the time.

The management of strabismus is to recognize the condition as early as possible and to remediate the underlying defect. Congenital cataracts and corneal opacities should be removed in the first few weeks of life, and surgery for crossed eyes should be carried out early, usually by age 2 or 3. Surgery can align the eyes so they look straight, but surgery does not directly influence amblyopia. Strabismus surgery in older children and adults can straighten the eyes so they appear normal cosmetically, but this surgery has no effect on vision.

The amblyopic eye is best capable of improvement during the critical period that ends at age 6–8, and the standard treatment is occlusion of the dominant eye by patching or by drops (which blur images falling on the normal eye), which forces the child to use the amblyopic eye. This means that it is critical for amblyopia to be diagnosed at a very young age—ideally before 3 years of age. If a child's amblyopia is diagnosed on a routine school vision examination, the child may be too old for effective amblyopia treatment. Once the critical period has passed, amblyopia often cannot be improved. On the other hand, if vision is compromised in an eye after the critical period, vision in the affected eye usually remains normal. Two clinical examples will suffice. The first is a teenage boy who sustained a contusion injury to one eye with development of a cataract that eliminated all useful vision. Yet 40 years later, when the cataract was removed, the eye quickly regained normal 20/20 vision. The second example is a 30-year-old woman who had strabismic amblyopia. She sustained a severe injury to her normal-seeing dominant eye in an auto accident, and it was enucleated. This provided the equivalent of full-time, complete patching. Yet in subsequent years there was essentially no improvement or change in the vision of that amblyopic eye. On the other hand, there are reported cases of patients in their teens or 20s regaining vision in an amblyopic eye after their nonamblyopic eye was damaged or lost. Little attention was paid to these reports for many years, but recently, concerted efforts are being made to improve vision in amblyopic eyes, and some of this research appears promising.

Amblyopia has been studied extensively in cats and monkeys, and these studies, initiated by Wiesel and Hubel in the early 1960s, have told us much about the underlying mechanisms. Two kinds of experiments have been undertaken—one in which sharp images are precluded from falling on the photoreceptors by putting an occluder over one eye in the young animal or by sewing the lids shut. In neither case is light excluded from the eye, but the images are blurred. In the second experiment the eyes are surgically crossed. The results obtained by the two procedures appear basically similar, but this has not been extensively studied.

The animal studies show, first of all, that at birth the cortex is wired up much as it is in the adult eye. Both eyes contribute equally to the cortical neurons, and responses recorded from the neurons are remarkably adult-like. Simple, complex, and specialized complex cells are recorded that require oriented stimuli presented to the retina. Further, most of the cells are binocular, receiving input from both eyes. Thus, the cortex, at least area V1, is basically wired up synaptically without needing any visual experience.

However with both cats and monkeys, monocular deprivation by an occluder or lid closure quickly causes visual defects.

Again, the deprivation must be done in the young animal, and dramatic changes are seen by examining the ocular dominance properties of the neurons in area V1. Whereas in the normal visual system most cortical cells in area V1 are binocular, with a deprived eye most of the cortical cells are driven only by the nondeprived eye and are, therefore, monocular. Figure 5.12 shows an ocular dominance histogram from a cat in which the lid of one eye was closed from the first week of its life to the fourteenth week. Recordings were then made from area V1. Virtually all of the cells were monocular, receiving input only from the nondeprived eye, and there were very few binocular cells. A number of cells encountered gave abnormal responses, and some cells gave no light responses at all. The conclusion drawn from these experiments is that as a result of the deprivation, the deprived eye no longer had the input to the cortex it initially had; in other words, there was likely a massive loss of synaptic connections in those parts of the cortex driven by the deprived eye.

A dramatic confirmation that the deprived eye loses representation in the cortex was shown subsequently in anatomical studies. Ocular dominance columns in layer 4 of the cortex can be visualized by injecting radioactive amino acids into one eye. The amino acids are taken up by the ganglion cells, synthesized into proteins, some of which are transported down the cell's axon to the LGN. There, some of the labeled protein is released, taken up by the LGN neurons and transported by their axons to layer 4 of the cortex. If a horizontal section is cut through layer 4 and the tissue section exposed to photographic film, wherever there is any radioactivity silver grains in the film will be exposed just as visible light exposes them when we take a photograph with film. The exposed area then appears black. Figure 5.13 shows representations of the ocular dominance columns as they run across layer 4 in a normal case and one eye deprived of sharp

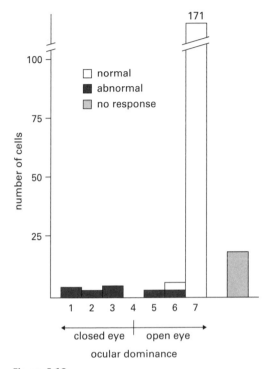

Figure 5.12
Ocular dominance histograms of cell responses recorded in the cortex of a cat that had one eyelid closed from the first to the fourteenth week of life. In the normal animal, relatively few cells have input from just one eye; the bulk of the cells have input from both eyes but to varying degrees (see figure 5.9). Here virtually all the cells had input from just one eye, the open eye.

images. In this case the radioactive amino acid was injected into the deprived eye. Whereas in the normal eye the stripes are of approximately equal size, indicating that each eye has equal representation on the cortex (equal-sized ocular dominance columns), this is not the case for the cortex from the deprived eye.

A   B

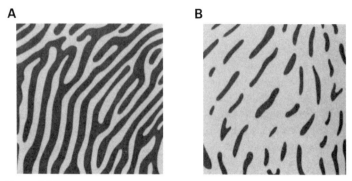

Figure 5.13

(A) A representation of the ocular dominance columns or stripes that extend across the primary visual cortex. The cells in one stripe receive their input preferentially from one eye, and the stripes alternate. Each eye has equal representation in the cortex in the normal animal. (B) In an animal in which form vision has been deprived in one eye, the amount of cortex receiving input from that eye is much reduced. The ocular dominance columns or stripes are thin and discontinuous.

The stripes from the injected deprived eye are thin and discontinuous, whereas the stripes from the normal eye are considerably wider and have filled in the discontinuities. In other words, input from the open eye has taken over much of the cortical space once occupied by input from the closed eye.

It is also possible to alter other aspects of the neuronal responses in the cortex. If, for example, an animal is raised during the critical period such that it sees only horizontal or vertical stripes, subsequent examination of both simple and complex cortical neurons shows that they have a definite bias for the orientations to which the eyes were exposed (figure 5.14). Other experiments showed that by raising animals under conditions in which they see little movement, or movement only in one direction, cortical cells are less movement sensitive or respond only to the direction of movement to which they were exposed.

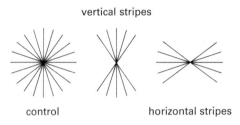

Figure 5.14

Representations of the orientation preferences for cortical cells in a control animal and in animals exposed only to vertical or horizontal stripes. In control animals all orientations are represented; in animals exposed only to horizontal or vertical stripes, the orientation preferences of the cells reflect the stimuli to which the eyes had been exposed.

As noted earlier, in wall-eyed (exotropic) individuals acuity is usually normal in each eye. When the individual looks to the right, he/she uses the right eye; when looking left, the left eye is used, but in both eyes vision is usually 20/20. These individuals do not have good depth perception because they have few if any normal binocular neurons in area V1. Figure 5.15 shows ocular dominance histograms in animals made cross-eyed or wall-eyed. In both cases the cells are mainly monocular, but in the "wall-eyed" animal, there are many cells receiving monocular input from one eye or the other, reflecting the fact that both eyes likely have good vision, whereas in the cross-eyed cat, the cortical cells mainly receive input only from one—the nondeprived—eye. In neither animal are there significant numbers of binocular cells. These results show that lost synapses can be quite specific; that is, synapses involved in binocular vision are lost, but not those involved in motion and direction sensitivity. Although less common, some strabismic patients also have an alternating esotropia in which each eye retains good vision.

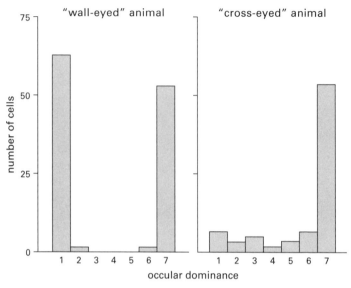

Figure 5.15
Ocular dominance histograms from animals made "wall-eyed" (left) or "cross-eyed" (right). In wall-eyed animals, most cells are monocular; in many cross-eyed animals, one eye becomes dominant, and most cells recorded are monocular, receiving input only from the dominant straight eye.

To conclude, most agree that after the age of 8 or so, amblyopia does not develop following visual deprivation, but relief from aspects of amblyopia can be achieved, at least partially, for many years thereafter by intense training of the amblyopic eye. On the other hand, recovery from amblyopia is faster and more successful in the early, critical period years. In Oliver Sacks's book, *The Mind's Eye* is the story of a woman, Sue Barry, who as a child had a crossed eye that was surgically straightened. She wore a patch and regained good vision in the eye but had no depth perception. As a result of visual exercises she regained some binocular vision and depth perception in her late 40s. It required intense work, and if she stops doing the exercises, her binocular vision regresses. But, her story makes it clear that the cortex can be modified with training probably all of our lives. This means neurons can grow new processes and form new synapses and/or perhaps reactivate synapses that have fallen silent. Clearly the young brain is more modifiable than is the older brain, and the notion of a critical period reflects a period of exceptional modifiability. But critical periods are not absolute—modification of brain structure and circuitry is possible throughout life. Indeed, that those in their 80s, 90s, or older can continue to learn and remember new things is evidence that brains remain modifiable for as long as we live.

6   Higher-Level Processing and Visual Perception:
Blindsight

Visual information from area V1 goes mainly to area V2 and from there spreads out into many cortical regions, perhaps as many as 30 or more. V2, adjacent to V1, has a stripe-like organization, as shown schematically in figure 6.1. First observed anatomically because the cells in the stripes stain more intensely for a particular enzyme, it was subsequently found by Margaret Livingstone and David Hubel in the 1980s that cells with different receptive field properties clustered in the stripes or between the stripes. Three patterns can be distinguished—thick and thin dark stripes and pale interstripe areas. In the thick stripes are found many direction-selective and highly binocular cells; in the thin dark stripes, there are many complex, orientation-selective cells that are end-stopped. Finally in the pale, interstripe areas, over half the cells are color coded with mainly double-opponent receptive fields that showed no orientation selectivity. The conclusion drawn from these results is that in V2, there is a further segregation of cells interested in form (thin stripes), movement (thick stripes), and color (interstripe areas). The segregation is not absolute—some color cells are found in the thin stripes, and some orientation-selective cells are found in the thick stripes.

Beyond area V2, there is further segregation of form, movement, and color processing, although again it is not absolute.

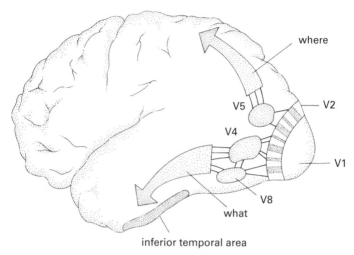

Figure 6.1

Visual areas beyond V1. Enzyme staining shows banding of area V2 into thick bands, thin bands, and interband regions, where different aspects of the visual image are processed. Beyond V2, separate areas are concerned with processing these aspects of the visual image: V4, form and color; V8, color; and V5, movement. Visual information progresses to yet higher visual areas via two pathways—a dorsal pathway into the parietal lobe known as the *where* pathway and a ventral pathway into the temporal lobe known as the *what* pathway. A face recognition area is localized on the underside and inner surfaces of the temporal lobes (inferior temporal area).

Area V5 seems to be primarily concerned with motion and depth, and if this area is damaged, monkeys have great difficulty in discriminating direction of movement. A few patients have been described who have lost the ability to see movement but have normal form and color vision, suggesting they may have damage in area V5. Loss of movement vision is debilitating; for example, such individuals cannot cross a street unaided because they cannot tell whether cars are moving toward them. They have difficulty pouring a cup of tea because the tea looks solid to them, and they cannot see the tea rising in the cup.

Loss of area V4 in a monkey impairs its ability to discriminate patterns and shapes, although color discrimination is only minimally affected. There are some patients with cortical lesions who have lost the ability to see colors but whose form and movement vision remain normal. This indicates that there is a color-specific area, comparable to V4 and V5, but debate continues as to exactly where it is. The guess is that it is ventral and anterior to V4, and some have labeled it V8 (figure 6.1).

A dramatic case of loss of all color vision, a condition, called achromatopsia, was described by Oliver Sacks, first in *The New York Review of Books* and then in his book *An Anthropologist on Mars*. Jonathan I. was a successful artist, 65 years of age, who created colorful abstract paintings. While driving in the city one day, his car was hit by a truck. Mr. I. seemed unhurt except for a severe headache. He slept deeply that night and the next morning could not remember much about the accident. He soon noticed, however, that he could not read—letters appeared like "Greek" to him, and he also discovered he could not see colors. He was tested at a nearby hospital that diagnosed him as having a concussion. His inability to read went away over the course of days, but his loss of color vision was permanent. This was devastating for Mr. I. His studio was filled with his brilliantly colored paintings, which now all appeared gray to him, and consequently the paintings had lost their meaning. He could see objects clearly, could make judgments of gray scales, and could see movement normally. He could read and draw accurately, but nothing had color any longer. Interestingly, he could not even visualize colors in his mind. He eventually did take up art again, but he drew strictly in black and white, and he converted to sculpture as his main artistic medium.

From areas V4, V5, and V8, visual information goes to numerous other areas along two general pathways. One pathway runs dorsally in the cortex toward its top (figure 6.1) and is known as the *where* pathway. It provides information about where an object is in space and is involved in the visual control of reaching and grasping an object. The *where* pathway receives much input from area V5 and some from areas V4 and V8.

The other pathway runs ventrally, toward the bottom of the cortex and is known as the *what* pathway—it is involved in object recognition. Patients with damage to the *what* pathway may have difficulty visualizing an object or identifying it, but they can reach out and grasp an object appropriately, even when its orientation is altered. This was demonstrated clearly in a patient whose *what* pathway was damaged by carbon monoxide poisoning. She could not identify a pencil shown to her but could grasp it accurately whether it was in a vertical or a horizontal position. She also could walk along a rocky path without tripping or running into objects such as tree limbs, but she did not know why she was making her evasive movements. Patients with lesions in the *where* pathway can see objects and identify them, but they have great difficulty in locating objects in space and grasping them. The *what* pathway gets substantial input from area V4 but also some from V5 and V8.

Object and Face Recognition

The further along the *what* pathway, the more specific are the cells in terms of what best activates them. Of particular fascination are cells found in a region of the cortex called the inferior temporal area. A number of cells there appear to be specialized for recognizing faces, as first discovered by Charles Gross in the 1970s. Clinical cases are reported in which individuals who have had strokes that affect the inferior temporal cortex have lost the ability to recognize people visually. These patients may be free of other visual deficits. They can read, write, and name objects, but they fail to recognize even their spouses entering the room. As soon as the spouse speaks, however, the patient will immediately recognize the voice and identify the person, showing clearly it is a visual defect. The patient sees that a person has entered the room and can even describe features of the person but cannot recognize his or her face.

The failure to recognize faces is termed *prosopagnosia*, and it is occasionally present at birth. It is estimated that about 2% of the population in the United States, about 6 million people, are affected with this disorder. Interestingly, Oliver Sacks, who has written about so many brain disorders, had this deficit and wrote extensively about it in his book *The Mind's Eye*.

His brother also has prosopagnosia, suggesting it has a genetic basis. People born with the disorder and others with it may also have trouble recognizing places, and they frequently get lost. Among other things, Sacks noted that he is better at recognizing dogs than people because dogs have distinctive sizes and shapes.

There are some people described by Ken Nakayama and his colleagues as "super-recognizers" who have supernormal face-recognition abilities. Some of these individuals claim to be able to recognize virtually every face they have ever seen. They may not know the person's name, but they remember where they encountered that person before. This ability holds for other aspects of object recognition. John's wife, who is an expert in Asian Art, has a remarkable ability to recognize pieces of art she has seen before and remembers where she saw them.

Can we say anything neurobiologically about face recognition? Recordings have been made from the inferior temporal region in monkeys, and certain neurons respond vigorously when the monkey is shown a monkey face (figure 6.2). If features

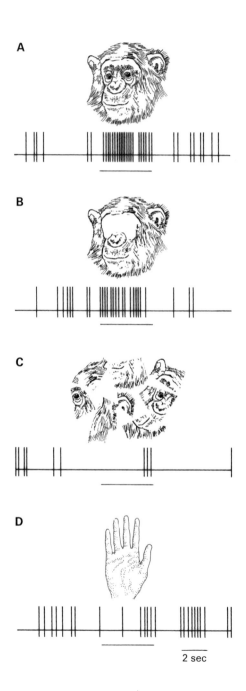

2 sec

Figure 6.2
Recordings from a monkey inferior temporal neuron that responded best to faces. The images were projected onto the retina for 3 seconds (bars below action potential records). The neuron responded best to the intact face (A), less well to an image in which the eyes have been blanked out (B), and not at all to a picture of a face cut into 16 pieces and rearranged (C) or to a hand (D).

of the face are eliminated, the neurons respond less well. For example, if the eyes are blanked out, the neurons respond less vigorously, and if the face is jumbled so that all the complexity of the face is present but inappropriately arranged, the neurons do not respond at all. And the neurons do not respond to a hand, but other neurons in the region will respond to hands and other complex objects selectively.

Face recognition has some curious aspects. For example, it is difficult to recognize a face when it is upside down. Indeed, if features of a face are distorted, a face looks much more normal upside down than right side up, as shown in figure 6.3. These images of a former well-known political figure look quite similar when viewed upside down, but when viewed right-side up, her face appears grossly different.

Visual Processing and Perception

Three general themes have emerged from the study of visual processing at all levels of the visual system, from the retina to visual areas V1, V2, V4, V5, V8, and beyond. These generalizations offer insights into how we perceive images and the nature of visual perception.

The first is that cells and pathways in the visual system are concerned with one or another aspect of the visual image. At the level of the photoreceptor synapse, ON and OFF information is

Figure 6.3
Upside-down pictures of Mrs. Thatcher. The two pictures appear approximately equivalent when viewed upside down, but when turned right
side up they appear very different. In one of the photographs, the eyes
and mouth have been inverted.

separated into two classes of bipolar cells, and this segregation
is maintained throughout the visual system. Further, the outer
plexiform layer of the retina is concerned with spatial aspects
of a visual image (and color to some extent), whereas the inner
plexiform layer is concerned more with the dynamic or temporal aspects of a visual scene (i.e., movement). Basic classes
of ganglion cells providing information about the processing
of visual information in the outer and inner plexiform layers
project to higher visual centers. Segregation of these aspects of
the visual image is maintained in the lateral geniculate nucleus
and throughout the cortex. Beyond visual areas V1 and V2, specific components of the visual image involving form, color, and
movement are dealt with separately and in different cortical
areas. There exist, then, parallel processing streams in the cortex
that give rise, ultimately, to the relatively unified visual world

we perceive. Massive parallel processing occurs throughout the brain, and our brains are multitasking all the time, carrying out many operations simultaneously—seeing, hearing, smelling, talking, walking, holding an object, and so forth.

A second important theme is that the visual system is not designed to make absolute judgments but rather to make comparisons. This characteristic is seen first in the outer retina and is reflected in the receptive field organization of bipolar and ganglion cells. That is, the receptive fields of bipolar and many ganglion cells consist of antagonistic center-surround or color-opponent regions. As noted in chapter 3, it is not the intensity of light that comes from an object that makes it look light or dark but the intensity of light coming from the object relative to the intensity of light coming from surrounding objects. This was shown dramatically in figure 4.3, and another example is shown in figure 6.4. Two simple examples further serve to illustrate this. When turned off, a television screen usually appears gray (at least older black-and-white TV sets do). When turned on, the television picture displays blacks as well as all shades of gray and bright whites. There is no such thing as negative light; the natural gray of the television screen appears black when the set is on because of adjacent bright areas. Reading a newspaper in very dim or bright light provides another example. In both cases the print appears black, and the rest of the paper white. The light reflecting off the black print in bright light can exceed the intensity of light reflecting off the white areas in dim light.

Not only does perceived brightness depend on surrounding illumination, but an object's perceived color does also. The color of an object can look quite different in surrounds of different color. No one selects upholstery material for curtains, a chair, or a sofa without taking a swatch home to see if it looks right in the designated space. This is shown impressively in plate 8. The Xs on each side of the figure look distinctly different in terms of

Figure 6.4
The bar is uniformly gray from one side to the other but appears lighter against a dark background, while a lighter background makes it appear to be darker.

color, but at the top of the figure where they come together, it is clear they are the same color on both sides. Surround is even critical for perceived judgments of size. Figure 6.5 shows pictures of two women along a corridor. The women appear of comparable size in the picture on the left, but the women in the right picture appear vastly different in size. However, the size of the image of the woman in the right picture is identical to her image in the left picture. Measure her yourself, but she appears tiny in

Figure 6.5
Context affects the perception of size: the scene was photographed twice: once with both women present, once with only the woman in the foreground. The image of the other woman was cut out from a print of the first picture and pasted onto the second picture. She now appears much smaller than in the first picture, but she is exactly the same size in both photographs.

the right-hand picture because the surrounding objects are not of the appropriate size relative to the size of her image.

The latter example brings us to the major and most important generalization about visual perception—that is, it is *reconstructive* and *creative*. For example, we live in a three-dimensional world, yet the images falling on our retinas are two dimensional. We construct the three dimensionality in our brains. Our visual system uses many pieces of information to construct an image. Not only is incoming information important, but also our experience and expectations are critical. If the incoming information is not consistent or is not complete, the visual system attempts nevertheless to provide a complete and coherent image.

The visual system can be confounded, however, and visual illusions illustrate this nicely. Figure 6.6 shows three familiar illusions. No lines in these illusions demarcate a triangle in the top illusion, a circle in the middle illusion, or the separation of the two sets of parallel lines in the bottom illusion. Yet we clearly "see" a triangle and a circle, and unless you look closely, there seems to be an actual line separating the two sets of lines in the bottom illustration. Further, the triangle and circle even look brighter than the surrounding areas that are equally white.

That visual perception is reconstructive and creative was appreciated at the beginning of the twentieth century by Gestalt psychologists, who put forward the idea that the brain analyzes an image by taking in information concerning components of a visual scene and by making assumptions about the shapes, colors, and movements within the scene. The brain then creates a coherent and consistent mental picture. If something is missing from the scene, it is filled in as is nicely demonstrated in the visual illusions of figure 6.6. Another good example is the fact that we have a blindspot in our retinas, where the optic nerve exits (see figure 4.11). But when we look out at the world we do not see a blind spot in our field of view. The brain fills in the defect so that we do not notice it. Yet, the blind spot can be easily demonstrated by the simple exercise of closing one's left eye and then focusing one's right eye on the X in figure 6.7. If you move the book toward or away from your face, the butterfly on the right disappears. Twist the book and it instantly reappears. It also reappears when you move the book closer to your face or further away.

If an image we are looking at is ambiguous, we perceive one thing or another but not a mixture of the two percepts. An example of this is the famous vase–face illusion shown in

Figure 6.6
Examples of illusionary borders. Although the shapes are not drawn as such, a well-defined triangle and circle appear in the top two figures. In the lowest figure, a curved line separating the two halves of the figure appears to be visible.

Figure 6.7

The physiological blind spot. Close your left eye and focus your right eye on the X. Move the book slowly toward or away your face and the butterfly on the right will eventually disappear. It reappears when the book is twisted or moved one way or the other.

figure 6.8. At any one instant in time we may see two black faces in the picture or a white vase, and it is easy to switch back and forth between the two percepts, but we see only one of the two percepts at a time.

The latter illusion illustrates another feature of visual perception, namely that attention is critical for extracting visual information from a scene. To switch from one percept to the other requires attention. If there are multiple objects in a visual scene, and we focus our attention on one or a few objects, perception of the other objects in the field of view is often completely ignored. An impressive demonstration of this is a video in which a basketball is being tossed from one person to another. The viewers are asked to focus initially on the tossing of the basketball and count how many times it is passed from one player to the next. During the video a man in a gorilla costume walks across the scene, but virtually no one "sees" him. When the video is replayed and the viewers are asked to look for the man in the gorilla suit, everyone sees him. Another physiological experiment demonstrating the role of attention is shown in figure 5.1. When a monkey attends to something in the periphery of its visual field, the activity of neurons that receive input from that area shows a significant increase in the number of action potentials generated.

Figure 6.8
The face–vase illusion. At any one time two black faces or a white vase is perceived. Never are the two percepts seen simultaneously, although some people can switch percepts so quickly, they think they can see them simultaneously.

How images are created in the cortex is not known. Interactions among the processing streams must occur, perhaps at several levels. This is known as the "binding" problem—how does the cortex assemble pieces of information about the form, color, depth, and movement in a scene, couple these with our visual memories and attention, and create a recognizable visual image? Several ideas have been suggested, but unambiguous evidence for any of these ideas is lacking. There does not appear to be one brain area that coordinates color, form, motion, and depth, and so the brain must use another strategy, but what that might be is not known. We construct coherent images in our brain, but not necessarily the ones that are out there. Other systems in the brain appear to operate in a similar fashion. Autobiographical memories are often reconstructive and creative, reflecting the confluence of a number of past experiences. Our visual system is not faultless, nor is any other part of the brain, but an important generalization is that our brains strive to give us a logical, consistent, and coherent view of the world.

Blindsight

With a large lesion in area V1, patients lose all visual perception and claim they are totally blind. They do, however, retain surprising visual abilities, and this phenomenon is called blindsight. Blindsight has been studied extensively by Lawrence Weiskanz and his colleagues beginning in the 1970s and since. For example, if such patients are asked to point toward a spot of light on a dark screen, they respond that they cannot see anything. When asked to guess where the spot is, they point directly to it. One such individual, when facing large moving stripes projected on a screen, followed the stripes with his eyes. When asked why his eyes were moving, he could provide no explanation. Pupils in these individuals respond to light, and such patients often have the ability to grasp objects correctly when asked to reach for something they are told is in front of them.

How might this phenomenon be mediated? Ganglion cells project to a number of subcortical areas in addition to the lateral geniculate nucleus and area V1. Indeed there are as many as nine such areas. These areas are concerned with a variety of visual phenomena such as initiating eye movements, regulation of circadian rhythms, pupil contraction, the ability to follow moving stripes (the optokinetic reflex), and so forth. One suggestion, and the most reasonable, is that blindsight phenomena are mediated by these subcortical areas. It has also been proposed that some of these subcortical areas project to higher cortical areas such as area V5, bypassing area V1, to explain this phenomenon. If this were so, one might expect that the patients would have some awareness of the visual stimuli presented to them, but they do not. It is generally believed that visual perception/awareness is a cortical phenomenon, and if cortical areas other than V1 are being activated, it would seem that individuals would have some recognition of what they are responding to, and they do not.

This discussion raises the interesting question of awareness and visual perception in animals that have little or no cortex. A frog, for example, can track a fly visually very well and capture it with its tongue if the fly is close enough. But does it "see" the fly as we see a fly? Is it aware of its presence? If a frog is put into a cage with flies buzzing around, it does fine, readily capturing them. However, if the frog is put into a cage with dead flies that are perfectly edible, it will starve to death. It is not aware of them—it doesn't "see" them. Why? Because they are not moving, and a frog's visual system is tuned to detect moving stimuli. This is also the case with other animals. At what level in the animal world, do animals experience visual perception much as we do? No one knows.

7   Looking Back and Forward

Philosophers, physicians, physicists, biologists, and others have long been interested in the phenomenon of vision. This chapter briefly reviews chronologically some of the major historical advances in our knowledge of the basics of vision. The ancient Greeks wondered how the eyes work. A major debate then was whether eyes send out rays to see what is out there or whether light rays from the outside enter the eye to produce sight.

This debate was not settled until the tenth century AD, and credit is given to a famous Arab scientist, Ibn Al Hatham (Alhazen), who carried out a simple experiment. He exposed animal eyes to the sun and found, after a period of time, the retina was damaged. He reasoned that light from the sun entering the eye caused the damage and that is how we perceive images: light rays enter the eye and are sensed there.

Little further knowledge was attained until the beginning of the nineteenth century, when there was an explosion of new facts and ideas. First, Thomas Young in 1802, building on Newton's discovery that white light could be divided by a prism into its color components, proposed that different colors of light are perceived because they interact with specific "particles" (cones) in the eye. He suggested that there are at least three particles, broadly tuned, and which absorb maximally red, green, and blue light, that account for color vision. Herman von Helmholtz later confirmed and expanded Young's findings, developing the trichromatic theory of color vision, which has been impressively confirmed experimentally. Helmholtz also invented the ophthalmoscope, the instrument that allows one to look into the back of the eye—the fundus—and examine the retina. This instrument has been a fundamental diagnostic tool for ophthalmology ever since (plate 5).

With the development of the compound light microscope, rods and cones were described in detail in the mid-nineteenth century by Max Schultze (figure 7.1). He also identified the three cellular layers of the retina and was the first to suggest that cones function in bright light and rods in dim light. In 1876 Franz Boll proposed that the reddish-purple color that he observed in

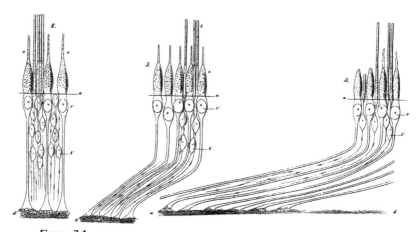

Figure 7.1
The 1866 drawings of Max Schultze of rods and cones in the human retina. The left is from the peripheral retina, and that on the right is from the central retina where the inner layers of the retina are swept aside (see figure 1.5). The middle drawing is in between.

a dark-adapted frog's retina, and which rapidly disappears in the light, was due to the existence of a light-sensitive pigment that initiates vision (plate 3). Wilhelm Friedrich Kühne quickly confirmed Boll's observations, and by studying the color changes that take place in a dark-adapted retina exposed to light, proposed a visual cycle, the basic outlines of which remain valid to this day (see figure 7.4).

In the late nineteenth century the great neuroanatomist Ramón y Cajal studied the retinas of many species using the silver impregnation method of Camillo Golgi. This method provided for the first time views of entire nerve cells, and it and similar methods are still in use today. Cajal was able to identify all the retinal cell classes, recognized that there are many subtypes of each cell class, and noted that there appear to be laminae in the inner plexiform layer that the processes of specific subtypes of cells run along (figure 7.2). He also studied the visual cortex and described the cells and the cellular layers there (figure 7.3) as well as the cells in most other parts of the brain.

In the beginning of the twentieth century, vitamins were discovered, and vitamin A was quickly shown to be essential for vision. Although evidence that a dietary factor is needed for vision dates back to the ancient Egyptian Medical Papyri, the specific factor was unknown. By the 1920s it had clearly been shown that nutritional night blindness was caused by vitamin A deficiency and that there is abundant vitamin A in the retina of a normal light-adapted animal.

George Wald, a postdoctoral fellow in Germany in the early 1930s, made the seminal discovery of the role of vitamin A in vision. Working with frogs, he found in accord with earlier work that when he extracted retinas with a mild organic solvent (petroleum ether) in the light-adapted state, he found abundant vitamin A. When dark-adapted retinas were first exposed to light, and had assumed a yellow color (Kühne's visual yellow,

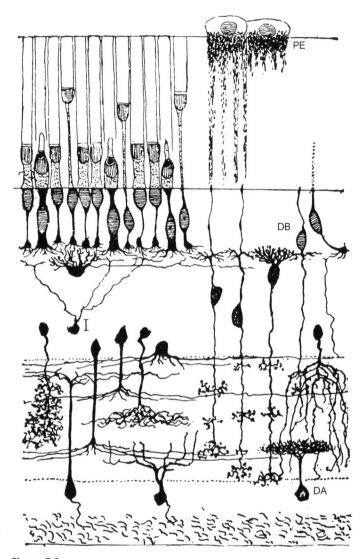

Figure 7.2
Drawing showing typical retinal cells of the frog and the cellular organization of the frog retina. PE, pigment epithelial cells; I, a partially stained cell; DB, displaced bipolar cell; DA, displaced amacrine cell. Modified from Ramón y Cajal (1911).

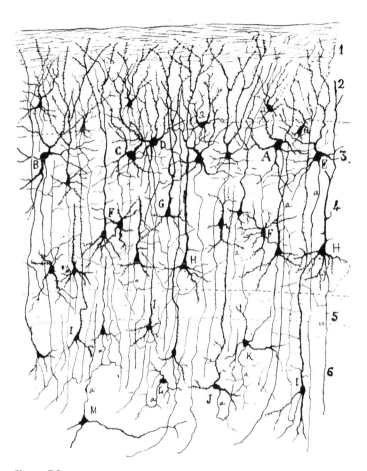

Figure 7.3
Neurons in the cortex drawn by Ramón y Cajal. Two types of cells can be recognized. The cells marked H are typical pyramidal cells; those marked F are typical stellate cells. The other cells are variant pyramidal and stellate cells. The layers of the cortex are indicated on the right (see figure 5.4).

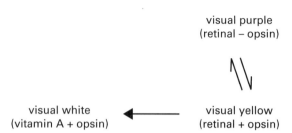

visual purple
(retinal – opsin)

visual white   ←————————   visual yellow
(vitamin A + opsin)   (retinal + opsin)

Figure 7.4
Kühne's visual cycle correlated with Wald's observations. Rod visual
pigment molecules consist of retinal (the aldehyde form of vitamin A)
bound to opsin (a protein) and appear reddish-purple in the dark-adapted state; hence the name "visual purple" for the native pigment (now
rhodopsin). Visual purple bleaches in the light to visual yellow; during
the bleaching process, retinal separates from opsin. Then the retinal is
converted to vitamin A, and the visual cycle reaches the visual white
stage (vitamin A and opsin). See plate 3.

figure 7.4), he extracted no vitamin A but a new substance that
had many properties of vitamin A but was clearly different—
it was yellow, for example, whereas vitamin A is colorless. He
named this new compound retinene (now called retinal). If he
extracted dark-adapted retinas with a mild organic solvent, he
extracted neither vitamin A nor retinal. However, if he used
a stronger solvent (chloroform) that denatures proteins and
releases molecules bound to a protein, he was able to extract
abundant retinal from dark-adapted retinas. He proposed, therefore, that visual pigments consist of a protein to which retinal is
bound and that light releases the retinal from the protein as well
as exciting the photoreceptor. This basic idea has been shown
to hold for all visual pigments of both rods and cones in every
species.

Also in the 1930s, strides were made on the physiological
front. Keffer Hartline recorded from single axons dissected from
the horseshoe crab eye (figure 7.5), which led to the discovery of
lateral inhibition in the distal reaches of the eye, a phenomenon
that has been found in every eye since and accounts for the fact

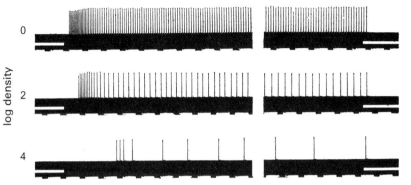

Figure 7.5
The electrical activity (discharge of action potentials) in a single optic nerve from the lateral eye of *Limulus*, the horseshoe crab, recorded by Keffer Hartline. Relative values of light intensity are given at left. Time marked in 0.2-second intervals in trace at bottom of each record; signal-marking period of steady illumination blackens out the white band just above the time marks. With brighter light stimuli more action potentials are generated.

that the brightness of an object depends on the brightness of the surrounding objects (figures 4.3 and 6.4). In vertebrates the lateral inhibition is mediated by the horizontal cells (chapter 4). Hartline also dissected out single ganglion cell axons from the optic nerve of the frog retina to show that some ganglion cells respond when the light goes on (ON-cells), others when the light goes off (OFF-cells), and some respond at both on and off (ON-OFF cells).

Microelectrodes that allowed researchers to insert an insulated sharpened wire close to neural cells were introduced in the 1940s and enabled investigators to record the receptive field properties of single ganglion cells from a wide variety of species. Landmark studies by Stephen Kuffler and Horace Barlow in the early 1950s showed that many of the receptive fields of ganglion cells were organized in a center-surround antagonistic fashion (see chapter 4), and Barlow showed that there are a variety of

ganglion cell types in frog and rabbit retinas, including directionally selective cells. David Hubel and Torsten Wiesel carried
the work into the cerebral cortex and made fundamental discoveries regarding how the cortical neurons, especially area V1 cells,
code visual information and how they are organized. They first
identified simple, complex, and specialized complex cells in area
V1 in the cortex and described the columnar organization of the
visual cortex. Wiesel and Hubel also pioneered studies on the
effects of visual deprivation on cortical neurons in young cats
and monkeys.

Intracellular micropipettes were introduced in the 1950s,
and enabled Gunner Svaetichin to make the first intracellular

Figure 7.6
Intracellular recordings from the mudpuppy retina showing the difference in response of a given cell type to a 100-µm spot and to a ring of
light (annulus) 1.0 mm in diameter, both at the same intensity. Note
that the distal retinal cells respond to light with sustained, graded, and
mainly hyperpolarizing responses. The receptor was probably a rod with
a relatively narrow receptive field, and the annular stimulus evoked very
little response. The horizontal cell, on the other hand, responded over
a much larger area, so relatively large hyperpolarizing responses were
recorded with both stimuli. The hyperpolarizing bipolar cell responded
by hyperpolarizing when the center of its receptive field was illuminated
(left column). With central illumination maintained (right column), annular illumination antagonized the sustained polarization elicited by
central illumination, and a response of opposite polarity was observed
in the cell. The transient amacrine cell gave depolarizing responses at
both onset and cessation of illumination. Its receptive field is somewhat
concentrically organized, giving a larger ON- response to spot illumination and a larger OFF- response to annular illumination of 1 mm diameter. The ON-center ganglion cell responded with a sustained depolarization and maintained discharge of action potentials to spot illumination.
With central illumination maintained, large annular illumination (right
column) inhibited impulse firing for the duration of the stimulus. The
ON-OFF ganglion cell gave bursts of impulses at both the onset and offset of illumination. Its receptive field was similar to that of the transient
amacrine cell. Modified from Werblin and Dowling (1969).

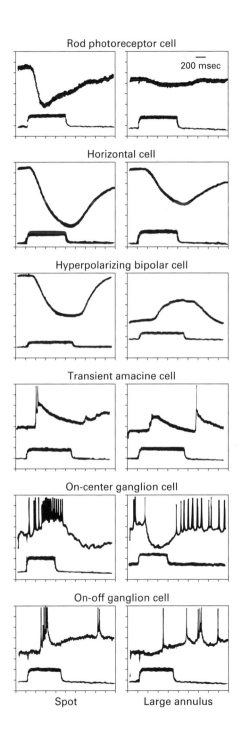

Rod photoreceptor cell

200 msec

Horizontal cell

Hyperpolarizing bipolar cell

Transient amacrine cell

On-center ganglion cell

On-off ganglion cell

Spot   Large annulus

recordings in the retina, first in horizontal cells, and then by Tsuneo Tomita in photoreceptors. These studies showed that these distal cells hyperpolarize in response to light. By the mid-1960s intracellular recordings were made from other retinal cells, and by the end of that decade, the responses of all retinal cell classes had been recorded by Werblin from the mudpuppy retina (figure 7.6).

The 1970s and 1980s revealed much about the pharmacology of the retina, showing that amino acids play a dominant role, with glutamate being the excitatory neurotransmitter released at both photoreceptor and bipolar cell synapses, and γ-aminobutyric acid (a slightly modified form of glutamate) and glycine, the major inhibitory substances used by the retinal interneurons. Further, dopamine, derived from the amino acid, tyrosine, was shown to play a major neuromodulatory role in the retina, serving to modify synaptic transmission at many retinal sites by strengthening or weakening synaptic efficacy of both chemical and electrical synapses. The 1980s also saw the elucidation of the photoreceptor transduction cascade, revealing the mechanisms underlying the hyperpolarization of photoreceptors in the light.

In 1990 Thaddeus Dryja and Eliot Berson identified a mutation in the rhodopsin gene in humans with retinitis pigmentosa (RP), and this set off an explosion of studies identifying mutations underlying all forms of RP and other ocular diseases. In the beginning of this century gene therapy for one form of RP was introduced, first in dogs and then in humans. Today genes are inserted into various species or knocked out (eliminated) or modified in various ways to increase understanding of both the normal and abnormal retina.

Dealing with Blindness

Despite much progress in the past 100 years or so in understanding vision and retinal disease, there are currently an estimated

1 million Americans who are presently legally blind. What are the prospects for future advances in the treatment of visual disabilities and blindness?

Most cases of blindness involve the eye, sometimes involving just one cell type—photoreceptors or ganglion cells, for example—but in other cases, the entire retina is involved. We have discussed gene therapy for Leber's congenital amaurosis, a retinal degeneration with similarities to other inherited retinal degenerations. This particular condition is the result of a single gene mutation, and the results of gene therapy in humans with this disease have been encouraging. Unfortunately the outlook for similar results in related retinal degenerations poses some complex problems. First of all, there are now 200–300 genes known that, when mutated, cause photoreceptor degeneration. For each retinal degenerative disease, a different viral vector containing the normal gene would need to be developed. Second, once a photoreceptor or pigment epithelial cell dies, gene therapy will not be effective. Thus, the virus vector must be presented to the retina when the affected cells are still alive. The question then arises as to when such therapy is indicated. Obviously, the earlier the better, but that would depend on the capability of diagnosing the disease and recognizing its prognosis with certainty at a very early stage.

Restoring light sensitivity to a retina that has lost its photoreceptors by injecting genes for light-sensitive molecules that are then expressed in bipolar or ganglion cells, an approach termed optogenetics, is another promising approach, but unless light sensitivity is restored to remaining and living photoreceptors, the acuity of the eye will be poor. This is because of the structure of the fovea where bipolar and ganglion cells are displaced laterally from the fovea's central region (figure 1.5). Images falling on photosensitive bipolar or ganglion cells in the central retina would be highly distorted and presumably not easy—or perhaps impossible—to interpret. Thus, making bipolar or ganglion cells photoreceptive would restore light sensitivity to the retina, but would the limited result be worthwhile? And if the ganglion cells or optic nerve are severely compromised in their function, visual information could not be transmitted to higher visual centers.

Restoring light sensitivity to remaining cones in a diseased retina may be more promising than making other retinal cells photosensitive. Cones usually survive longer than rods in most forms of retinitis pigmentosa and other diseases, and if light sensitivity could be restored and the cells kept alive, this could be effective therapy.

Another approach receiving much attention is the development of visual prostheses that activate ganglion cells or cortical cells directly. Two kinds of prostheses are being developed. One type replaces much of the retina and sits either inside or on the surface of the retina and thereby activates either bipolar or ganglion cells. The other class of prosthetic device bypasses the retina and activates cortical cells directly. In both situations a photoreceptive device is attached to the head or to glasses and feeds into electrodes implanted in the retina or cortex. Two retinal prosthetic devices are currently approved for human use, but they have a way to go before they provide reasonably good vision.

Yet another approach to replace damaged or destroyed photoreceptor cells is to transplant healthy photoreceptor cells into the eyes of blind animals. This strategy has had limited success so far—the number of transplanted cells that survive and integrate into the retinal circuitry is quite limited, but some restoration of electrical activity recorded from the eyes occurs, and the animals do show some behavioral responses to light. Stem cells, which in theory can differentiate into any cell type, have also been introduced into blind eyes, including some human eyes, again with very limited and largely undocumented success. Investigators are now inducing stem cells maintained in culture to differentiate into photoreceptor cells and then are injecting such cells into eyes whose photoreceptor cells have degenerated; this approach appears promising and may be more successful.

In addition to direct deleterious effects of a disease process or gene defect in the photoreceptor cells themselves, similar defects can also occur in the associated retinal pigment epithelial cells, and this can cause photoreceptor death. As discussed, the photoreceptor cells and overlying retinal pigment epithelium are intimately connected, and they depend on each other to function. The isomerization of vitamin A to generate 11-*cis*-retinal, needed to regenerate the visual pigment molecules after light exposure, occurs mainly in the retinal pigment epithelium, as does the phagocytosis and digestion of spent outer segment material. Compromise of these retinal pigment epithelial cell functions results in photoreceptor cell degeneration in both animals and humans. Thus, gene therapy to correct retinal pigment epithelial cell defects or transplantation of retinal pigment epithelial cells into diseased retinas has been accomplished with promising results. Indeed, the first gene therapy treatment in humans, described above, was for a gene defect in the retinal pigment epithelial cells. Further, retinal pigment epithelial cells grow readily in culture and are readily transplanted. Unlike photoreceptor cells, they do not need to integrate into the retinal circuitry but interact only with the photoreceptor cells, which they do readily.

It has long been known that nonmammalian species such as amphibians and fish can regenerate retinal cells endogenously, but mammals, including humans, cannot. Why can these cold-blooded vertebrates do this? This is an intriguing question that is now receiving substantial attention. If we could regenerate our retinal cells, presumably we could cure not only blindness caused by photoreceptor degeneration but blindness caused by degeneration of any retinal cell including the ganglion cells. In fish, for example, new neurons are formed throughout life, and the axons of the newly formed ganglion cells extend into the rest of the brain and make appropriate connections. In mammals, not only do ganglion cells not regenerate, but their axons do not regrow in large numbers after the optic nerve is damaged or cut.

From what cells does the regeneration in the nonmammalian species occur? This may differ among species, but certainly retinal pigment epithelial cells and Müller glial cells appear to be involved. In fish the formation of new retinal cells throughout life comes from a region in the retinal periphery called the marginal zone, whose cells may derive from the retinal pigment epithelium, whereas when the fish retina is damaged, new retinal neurons derive from Müller cells that dedifferentiate and appear to behave like stem cells. That is, after dedifferentiation these cells first proliferate and then generate progenitor cells for repairing the retina.

This discussion emphasizes some of the difficulties scientists face in seeking major breakthroughs in curing various causes of blindness. But make no mistake: thousands of dedicated scientists worldwide are attacking these problems, and we have no doubt that amazing developments will soon appear for the cure of various causes of blindness and to the benefit of mankind.

The Future

We have come a long way in understanding how we see and what can go wrong with the visual system, but there is still much to learn. Whereas the retina may be the best-understood part of the brain, the details of retinal circuitry underlying the receptive field organization of most of the subtypes of ganglion cells remain to be elucidated. Furthermore, we know that the properties of the retinal ganglion cells are altered after changes in illumination or as a result of a variety of other factors including fast eye movements or stimuli presented peripherally in the retina, well outside the classic receptive fields of the ganglion cells. In other words, the circuitry underlying the responses of retinal neurons can be modified significantly by such contextural stimuli—it is not static—and these modifications mainly represent the strengthening or weakening of synapses. This is likely the result mainly of neuromodulatory synaptic effects, but we still know little of the exact mechanisms or their influence on visual responses. Some of the substances inducing these neuromodulatory effects are known, such as dopamine, but there are no doubt many more. For example, we know that as many as 15 small neuropeptides are found in retinal neurons, especially amacrine cells, and they are presumably released at retinal synapses, but we know almost nothing about their individual functions.

In the cortex we are just beginning to understand the circuitry resulting in the receptive field properties and organization of the various cortical neurons and the underlying neuronal mechanisms. The simple schemes shown in figures 5.7 and 5.8 suggesting how simple and complex receptive field organizations may be formed were proposed 50 years ago, and we have not discovered much more since on this problem. There is still much to learn.

The Holy Grail in terms of cortical mechanisms is to understand visual perception. Features of the visual image are analyzed quite separately in different visual areas, yet they all come together when we perceive a scene or object. There is no evidence for a specific area in the cortex in which this happens, leaving us with a huge hole in our understanding of what many call the "binding problem." How does information about form, color, movement, and depth come together to give us the richness of visual perception? Insights into this formidable problem are likely to shed light on the greatest mystery in neuroscience—the mechanisms underlying consciousness.

Clinical Questions

There is still much to be known about the major eye diseases causing blindness, namely cataracts, macular and other retinal degenerations, glaucoma, and diabetic retinopathy. We have presented a consensus of the current state of knowledge of the underlying causes of these diseases, but these views are not set in stone and are subject to change as more research and experience accumulate. It is evident that studying disease in human beings is difficult for many reasons. Most importantly every human being is different (even identical twins) because of genetics and environment, and so in various individuals, the same diseases may be manifest differently, leading to many different notions of disease etiology (cause). A second reason is that it is difficult to analyze a disease in humans—one is limited in what can be done. For example, a patient can be observed, perhaps small tissue samples examined, but invasive technologies that might be useful for understanding a disease cannot be undertaken. Post-mortem autopsy studies provide important information but are not the same as studying a living organism. Further, every patient is unique, and so the sample size is essentially one for a particular disease. One way to deal with these problems is to induce a particular disease in susceptible animals. But animals can show quite different disease characteristics as compared to humans, and even different inbred strains of animals often demonstrate different disease patterns.

Despite the difficulties of studying human diseases, this is not to say that progress has not been made. Indeed, although there are disagreements, enormous advancements are being made, and new ideas and approaches to alleviating eye diseases are regularly accomplished. New techniques are being developed all the time that are giving the clinician new methods of viewing ocular structures in ways thought inconceivable not so many years ago. Optical coherence tomography (OCT), discussed and illustrated briefly in chapter 3, allows for imaging the retina in cross section in the living human eye much as a histological section of the retina does. Although the resolution of OCT is not yet comparable to that of a high-quality histological section (i.e., figure 1.8A), improvements in the technique are continually being made.

Other emerging techniques are promising to provide images at the single-cell level in the living eye. Adaptive optics originally developed by astronomers to improve the resolution of telescopes by the reduction or elimination of optical aberration caused by light passing through the atmosphere, has been modified, especially by David Williams and his colleagues at the University of Rochester, to visualize the arrangement of photoreceptors all over the human eye, including the thin foveal cones. It does this by reducing the optical aberrations caused by light passing through the ocular media. Plate 9 shows the arrangement of red-, green-, and blue-sensitive cones in the fovea. The cones appear randomly arranged, although some clumping of the green cones occurs. As has been known for some time, most humans have many more red than green cones and very few blue cones, especially in the fovea. So far adaptive optics has been used mainly for basic science studies, but it is beginning to find its way into the clinic, for example, to determine the fate of cone cells in certain macular degenerative diseases.

With regard to ways of dealing with problems involving the cerebral cortex, the realization that the cortex is much more modifiable than previously believed has spawned a variety of approaches to deal with issues previously considered untreatable. The obvious example is the amblyopic eye. Many believe that vision can be restored from these eyes, and a variety of techniques are being tried. Can rewiring of the cortex occur in other situations where regions of the visual cortex have been damaged? We do not as yet know, but it is a possibility to consider.

And finally, for those who are blind for whatever reason, can useful vision ever be restored? The ultimate cures for blindness will also vary with the particular cause of the blindness. However, we are confident progress will continue to be made. We have come an enormous way since the time of the ancient Greeks, and the pace of discovery is accelerating exponentially. We live in exciting and hopeful times.

Glossary

Accommodation   The ability of the eye to focus near and far images on the photoreceptors, accomplished by shape changes of the lens.

Achromatopsia   The loss of all color vision, usually as a result of a cortical lesion.

Action potentials   The transient, all-or-nothing electrical signals that travel down axons carrying the output information of neurons.

Adenosine triphosphate (ATP)   An energy-rich molecule made mainly in mitochondria that powers biochemical reactions in cells.

Adenylyl cyclase   The enzyme that converts ATP to cyclic AMP.

Amacrine cell   An inner retinal interneuron. Amacrine cells are involved in mediating movement among other things.

Amblyopia   Poor visual acuity because of visual (form) deprivation in a structurally normal eye in a young animal or human. Commonly caused by crossed eyes in the young human.

Amino acids   The molecules that, when strung together, form proteins.

γ-Aminobutyric acid (GABA)   An inhibitory neurotransmitter in the brain.

Aqueous humor   Watery fluid in the anterior chamber that is responsible for maintaining intraocular pressure (IOP).

Areas V1–5   Areas in the occipital cortex concerned with processing visual information.

Astigmatism   Refractive error caused by variations in curvature of the cornea or occasionally the lens, thereby preventing a clear focus.

Astrocyte   A glial cell found in the ganglion cell layer and optic nerve.

Axon   Thin cellular branch that extends from a neuron to contact another neuron or effector cell (muscle fiber). Axons usually carry the output message from a neuron via action potentials.

Binocular cell   A cortical cell that receives synaptic input from both eyes.

Bipolar cell   A retinal neuron that carries the visual signal from the outer to inner retina.

Blindsight   The visual abilities of patients who have no visual perception and say they are completely blind.

Blind spot   The region of the retina where the optic nerve exits and where there are no photoreceptors.

Calcium (Ca^{2+})   A positively charged ion important in photoreceptor adaptation and in synaptic transmission.

Cataract   Cloudy or opaque lens.

Catecholamine   A class of monoamine derived from the amino acid tyrosine. Dopamine, norepinephrine, and epinephrine are examples.

Central nervous system   The brain and spinal cord.

Cerebral cortex   A 2-mm-thick layer of cells that covers much of the brain. Highly infolded in man, the cortex is divided into two hemispheres, which are further subdivided into four lobes, frontal, parietal, occipital, and temporal.

Channel A membrane protein that allows ions to cross the cell membrane. Channels are usually closed until activated by a specific stimulus.

Chloride (Cl⁻) A negatively charged ion primarily involved in the inhibition of neurons.

Choroid A bed of blood vessels situated between the sclera and pigment epithelium that provides oxygen and nutrients to the outer retinal cells, especially the photoreceptors.

Circadian rhythm Endogenous rhythm that regulates various bodily functions over a 24-hour period.

Color-opponent cell A neuron whose center and surround receptive field responses are sensitive to different colors.

Complex receptive fields Receptive fields of neurons recorded in visual areas of the cerebral cortex that respond best to oriented light or dark bars moving across the retina.

Cone photoreceptors The photoreceptors responsible for daylight and color vision. Three types of cones exist in the human retina that are most sensitive to red, green, and blue light.

Corneal endothelium The innermost cells of the cornea, responsible for keeping the corneal stromal layer dehydrated and transparent.

Corneal epithelium The outermost cells of the cornea, providing a protective barrier for the cornea.

Cortical columns Columns of neurons that run through and across the cortex and share similar properties—orientation or ocular dominance.

Critical period The period of time during development when an animal is particularly sensitive to altered environmental conditions.

Crystallins Proteins in lens cells that do not scatter light and provide for good transparency.

Current A measure of the number of electrons flowing through a wire, or ions crossing a membrane, per unit of time.

Cyclic AMP   A second-messenger molecule formed by the enzyme adenylyl cyclase from adenosine triphosphate (ATP).

Cytoplasm   The substances inside cells exclusive of the nucleus.

Dark adaptation   The time required for the eye in the dark to regain full sensitivity after light exposure.

Dendrites   Branch-like structures that extend from the cell body of a neuron and receive synaptic input to the cell.

Depolarization   The process by which the interior of a cell becomes more positive.

Detached retina   A clinical condition caused by separation of the retina from the pigment epithelium.

Diabetic retinopathy   Damage to the retina caused by diabetes.

Diplopia   Double vision.

Direction-selective cell   A cell in the visual system that responds selectively to a spot or bar of light moving in a particular direction across the retina.

Disparity-tuned cells   Neurons that respond only to stimuli precisely positioned within their receptive fields. Thought to be critical for depth perception.

Dopamine   A neuromodulator released from brain synapses that has been associated with two diseases, Parkinson disease and schizophrenia. It plays an important role in the retina, mainly by altering synaptic strength.

Dry eye   The condition in which the quantity and/or quality of the tear film is deficient, causing discomfort of the cornea.

Ectoderm   Cells on the outer surface of the embryo that become skin.

Electrical synapse   Junctions (channels) that allow ions to flow directly from one cell to the next.

Electrons   The small negatively charged particles that surround the nucleus of an atom.

Endoderm Cells lining the inside of an embryo that form the gut and other internal organs.

Esotropia A type of strabismus in which one eye is turned in; often referred to as crossed eyes.

Excitatory synapse A synapse that excites a neuron.

Exotropia The strabismic condition when one or both eyes are turned out. Such individuals are often referred to as "wall-eyed."

Fovea The central indented region of the retina that mediates high-acuity vision.

Ganglion cells The third-order cells in the retina whose axons form the optic nerve.

Glaucoma A blinding disease primarily of the retinal ganglion cells and optic nerve, often associated with increased intraocular pressure.

Glia Supporting cells in the brain that help maintain neurons, regulate the environment, and form the myelin around axons.

Glutamate An amino acid that serves as the major excitatory neurotransmitter in the retina and rest of the brain.

Glycine An amino acid that serves as an inhibitory neurotransmitter in the retina and brain.

G-protein A protein activated by postsynaptic membrane receptors. Usually linked to an enzyme that makes a second-messenger molecule.

Graded potential A voltage change across a cell membrane that varies with the strength of the stimulus eliciting it. With a stronger stimulus, a larger potential occurs, and vice versa.

Growth factors Small proteins important for cell growth, differentiation, and survival.

Guanylyl cyclase The enzyme that makes cyclic GMP.

Histology The microscopic study of tissues.

Horizontal cell   An outer retinal interneuron that mediates lateral inhibition between photoreceptors and between photoreceptors and bipolar cells.

Hypercolumn   The basic module of the visual cortex. A piece of cortex 1 mm × 1 mm × 2 mm that contains all the machinery needed to analyze a bit of visual space.

Hyperopia   The refractive error in which objects are destined to focus behind the retina.

Hyperpolarization   The process by which the interior of a cell becomes more negative.

Inhibitory synapse   A synapse that inhibits neurons.

Inner plexiform layer (IPL)   The retinal synaptic layer where bipolar cells contact amacrine and ganglion cells.

Invertebrates   Lower animals such as insects, crabs, and mollusks that do not have a backbone.

Ion   An atom that is charged—that is, has gained an extra electron and is thus negatively charged or has lost one or two electrons and is thus positively charged.

Isomers (cis/trans)   Molecules made up of the same atoms but that have different shapes.

Kinases   Enzymes that add phosphate groups to proteins, thereby altering their function.

LASIK   A surgical technique that changes the cornea curvature, used to treat myopia and other refractive errors.

Lateral geniculate nucleus   The nucleus in the thalamus that receives input from the eye and transmits the visual signals to the cerebral cortex.

Lateral inhibition   The inhibition of one neuron by another, often reciprocal.

Lens nucleus   The central and oldest cells in the lens.

Light adaptation   The loss of visual sensitivity that occurs in the light.

Lysosomes Intracellular organelles whose role is to digest waste material.

Membrane (cell) The thin barrier surrounding a cell that keeps various substances in and other substances out. Consists of a lipid bilayer in which are embedded various kinds of proteins, including channels, pumps, enzymes, and receptors.

Mesoderm Cells between the ectoderm and endoderm in the embryo that develop into muscle, bone, and heart cells. In the early embryo, mesodermal cells induce overlying ectodermal cells to become neural plate cells.

Mitochondria Structures found in cells that provide the energy-rich molecules (i.e., adenosine triphosphate–ATP) that power the cell.

Modifying genes Genes that alter the effectiveness of other genes to express proteins.

Monoamines Substances released at synapses that function mainly as neuromodulators.

Müller cells The principal glial cell in the retina that extends through the retinal thickness.

Myelin Insulating layers of membrane formed around axons by glial cells.

Myopia A refractive error in which objects focus in front of the retina and distant vision is blurred.

Neural crest Cells found on either side of the neural tube in an embryo that form the peripheral nervous system.

Neural tube An early structure formed during brain development. Formed by the infolding of neural plate cells.

Neuromodulator Substance released at a synapse that causes biochemical changes in a postsynaptic neuron, thus modulating its function.

Neurons Cells in the brain involved in the reception, integration, and transmission of signals.

Neuropeptides Small proteins (peptides) released at synapses that act mainly as neuromodulators.

Neurotransmitter Substance released at a synapse that causes fast electrical excitation or inhibition of a neuron.

Night blindness Loss of light sensitivity at night resulting from impaired rod function. Can be caused by a deficiency of vitamin A or degenerative eye diseases.

Nucleic acid (DNA) The genetic material found in the nucleus of a cell that codes for the proteins made by the cell.

Occipital lobe The most posterior portion of the cerebral cortex, concerned with visual processing.

Ocular dominance The phenomenon that results in most binocular cells in the cortex being driven more strongly by one eye than the other.

OFF-cell A neuron whose activity decreases in response to light stimulation. Often at the cessation of the light such cells become more active for a short period.

ON-cell A neuron whose activity increases in response to light stimulation.

ON-OFF cell A neuron whose activity increases at the onset of a light stimulus and again at cessation of the stimulus.

Ophthalmoscope An instrument that allows one to look into the eye and view the structures in the back of the eye, the fundus.

Outer plexiform layer (OPL) The synaptic layer where photoreceptors contact bipolar and horizontal cells.

Outer segment The distal portion of photoreceptor cells where the visual pigments are found.

Parallel processing The simultaneous processing of information along separate pathways.

Peptide A small protein.

Phosphodiesterase An enzyme that breaks down cyclic nucleotides such as cyclic GMP or cyclic AMP.

Phosphorylation   The addition of a phosphate group to a protein. Serves to modify the properties of the protein.

Photon   The elementary unit (particle) of light.

Pigment epithelium   Pigmented cells that line the back of the eye and are in close association with the retinal photoreceptors.

Postsynaptic   Pertaining to structures or processes downstream of a synapse, as in postsynaptic neuron, postsynaptic potential.

Postsynaptic membrane   That region of a cell membrane specialized to receive synaptic input.

Potassium (K⁺)   A positively charged ion primarily involved in establishing the resting potential of a neuron.

Potential   Another term for voltage.

Presbyopia   The decreasing ability with age to focus on small near objects.

Presynaptic   Pertaining to structures upstream of a synapse, as in presynaptic neuron, presynaptic terminal.

Promoter gene   A gene that induces other genes to produce (express) a protein.

Prosopagnosia   The inability to recognize familiar faces.

Protein   A chain of amino acids folded in complex ways that enable the molecule to carry out its prescribed function.

Protons   The positively charged particles found in the center (nucleus) of an atom.

Pump   A membrane protein that moves ions across the membrane of a cell. Pumps require energy to function.

Pyramidal cell   A prominent neuron found in all areas of the cerebral cortex.

Receptive field   The area of the retina that, when stimulated, causes a visual system neuron to alter its activity.

Receptor potential   The voltage change elicited in a sensory cell or neuron following the presentation of a specific sensory stimulus to the cell or neuron.

Receptors   Membrane proteins found in postsynaptic membranes that are usually linked to intracellular enzyme systems; also, cells that respond to specific sensory stimuli, such as photoreceptors.

Refractive error   Impaired vision when images are not focused sharply on the photoreceptors.

Resting potential   The voltage across a cell membrane in the absence of any stimulus to the cell, typically –60 to –70 mV.

Reticular formation   Neurons found in the brainstem that extend widely throughout the brain and are important for regulating states of arousal, consciousness, and attention.

Retinal   The aldehyde form of vitamin A that, when combined with a specific protein, forms a visual pigment molecule.

Retinene   The original name for vitamin A aldehyde, now called retinal.

Retinitis pigmentosa   An inherited retinal disease that causes blindness.

Retinopathy   A general term referring to retinal disease.

Rhodopsin   The visual pigment of rods.

Ribosomes   Particles found in cells that are responsible for making proteins.

Rod photoreceptors   The photoreceptors responsible for dim-light vision.

Sclera   The tough white outer layer of the eye.

Second messenger   A small molecule synthesized in a cell in response to a neuromodulator (the first messenger).

Simple receptive fields   Receptive fields of neurons recorded in the primary visual area (V1) of the cortex that respond best to oriented bars of light or edges projected onto the retina.

Sodium (Na^+)   A positively charged ion involved in the generation of action potentials and in the excitation of neurons and sensory cells.

Specialized complex receptive fields Receptive fields of cortical neurons that respond to direction of movement or show other specialized features.

Stellate cell An interneuron found in the cerebral cortex.

Strabismus An eye muscle imbalance.

Synapse The site of functional contact between two neurons or a neuron and muscle cell.

Synaptic potential The voltage change produced in a neuron following the activation of a synapse impinging on the cell.

Synaptic terminal The site where a synapse is made.

Synaptic vesicles Small vesicles found at synapses that contain the chemicals released at the synapse.

Tear film The secretions from glands found underneath or on the eyelids, mainly watery, that help maintain the cornea.

Tectum A midbrain structure, especially prominent in nonmammalian species, that integrates sensory inputs and initiates motor outputs.

Thalamus A forebrain region that relays sensory information to the cerebral cortex.

Transcription factors Proteins that turn on and off the expression of genes by binding directly to the genetic material (DNA).

Transducin A G-protein found in photoreceptors linking light-activated visual pigment molecules with phosphodiesterase (PDE).

Trichromatic theory The theory that color vision is based on three color receptors that are differentially sensitive to red (long), green (middle), and blue (short) wavelengths of light.

Trophic factor A protein that encourages the growth of cells and/or protects cells from degeneration.

Ultraviolet light Short-wavelength light (below 400 nm) that is damaging to tissues.

Vascular endothelial growth factor (VEGF) A protein that stimulates the formation of new blood vessels.

Visible light   The wavelength of light humans can see—ranging from 400 nm (deep blue) to 700 nm (far red).

Visual pigments   The molecules in photoreceptors that absorb light and lead to the excitation of the cell and vision.

Visual prosthesis   An artificial device that replaces the retina and directly stimulates retinal cells (retinal prosthesis) or cortical cells (cortical prosthesis).

Vitreous humor   Gel that fills the back of the eye and serves to hold the retina in place.

Voltage   A measure of the electrical potential difference between two points; in neurons, the charge difference across the cell membrane (i.e., membrane voltage or potential).

What pathway   The ventral cortical pathway that is involved in object recognition.

Where pathway   The dorsal cortical pathway that is involved in spatial visual tasks.

White matter   Regions of the brain and spinal cord where there are abundant myelinated axons and relatively few cell bodies. The myelin gives the tissue its whitish appearance.

X chromosome   The so-called sex chromosome. Males have just one, but females have two.

Further Reading

Books and Reports

Astrocytes and Glaucomatous Neurodegeneration: A Report by the Lasker/IRRF Initiative for Innovation in Vision Science. 2010. New York: The Lasker Foundation. (One of a series of reports on major eye diseases, our understanding of them, and the major questions that need to be answered regarding them.)

Cajal, Ramon y, S. 1989. *Recollections of My Life.* Cambridge, MA: MIT Press. (A fascinating and very readable autobiography by the man many believe is the father of neuroscience. Some interesting science in it too.)

Daw. N. 2012. *How Vision Works: Behind What We See.* Oxford: Oxford University Press. (A detailed, extensively illustrated book on higher visual system function.)

Diabetic Retinopathy: Where We Are and a Path to Progress. A Report by the Lasker/IRRF Initiative for Innovation in Vision Science. 2012. New York: The Lasker Foundation. (The second in the series of reports on major eye diseases, our understanding of them, and the major questions that need to be answered regarding them.)

Dowling, J. E. 2000. *Creating Mind: How the Brain Works.* New York: Norton (paperback edition). (A gentle introduction on how the brain works. Requires little background in science.)

Dowling, J. E. 2012. *The Retina: An Approachable Part of the Brain* (Rev. ed.). Cambridge, MA: Belknap Harvard University Press. (A detailed overview of retinal structure, function, and pharmacology.)

Goodale, M. A., & Milner, D. 2005. *Sight Unseen: An Exploration of Conscious and Unconscious Vision.* Oxford: Oxford University Press. (A fascinating account of a woman poisoned by carbon monoxide who lost the ability to recognize objects but retained surprising visual abilities.)

Gregory, R. 1997. *Eye and Brain: The Psychology of Seeing.* Princeton, NJ: Princeton University Press. (A classic book introducing readers to the psychology of vision. First published in 1966, it has been through five editions but still has much valuable information in it.)

Herrman, D. 1998. *Helen Keller: A Life.* New York: Alfred A. Knopf. (A wonderful, detailed, and readable biography of Helen Keller.)

Hubel, D. H., & Wiesel, T. N. 2005. *Brain and Visual Perception: The Story of a 25-Year Collaboration.* Oxford: Oxford University Press. (An overview of the work of Hubel and Wiesel on the classic studies on the visual cortex for which they won a Nobel Prize. It begins with charming biographies of the two followed by annotated reprintings of many of their classic papers.)

Levin, L. A., Nilsson, S. F. E., & VerHove, J., eds. 2011. *Adler's Physiology of the Eye* (11th Ed.). New York: Elsevier. (A detailed textbook covering all aspects of eye structure and function with a clinical emphasis.)

Piatigorsky, J. 2007. *Gene Sharing and Evolution: The Diversity of Protein Functions.* Cambridge, MA: Harvard University Press. (An overview of gene sharing and its implications for evolution, starting with the lens crystallins.)

Polyak, S. 1941. *The Retina*. Chicago, University of Chicago Press. (The classic book on the anatomy of the primate retina. Very detailed but worth browsing for the illustrations.)

Ratliff, F. (Ed). 1974. *Studies on Excitation and Inhibition in the Retina: A Collection of Papers from the Laboratories of H. Keffer Hartline.* New York: Rockefeller University Press. (An annotated collection of original papers by one of the early visual electrophysiologists. Most of Hartline's work was on the horseshoe crabs, but he also was the first to record single ganglion cell axon responses from the frog retina, identifying ON-, OFF-, and ON-OFF ganglion cells.)

Restoring Vision to the Blind. A Report by the Lasker/IRRF Initiative for Innovation in Vision Science. 2014. New York: The Lasker Foundation. See also http://tvst.arvojournals.org/article .aspx?articleid=2212970/. (The third in the series of Lasker/IRRF reports that focuses on the various approaches being undertaken to restore vision to the blind.)

Sacks, O. 1995. *An Anthropologist on Mars*. New York: Alfred A. Knopf. (A series of case reports about individuals who have neurological disorders, including one of an artist who lost the ability to see colors.)

Sacks, O. 2010. *The Mind's Eye*. New York: Alfred A. Knopf. (A very readable book on vision and a variety of interesting visual disorders including the inability to recognize faces.)

Snyder, C. 1967. *Our Ophthalmic Heritage*. Boston: Little, Brown. (A fascinating collection of articles on historical figures and events in ophthalmology and vision research, including a description of the first successful corneal transplant by Edward

Zirm in 1905, and cataract surgery by couching described by Aurelius Celsus in 29 AD.)

Stone, J. V. 2012. *Vision and Brain: How We Perceive the World.* Cambridge, MA: MIT Press. (An introduction to vision science emphasizing computational approaches but without mathematical details.)

Weiskranz, L. 1986. *Blindsight.* Oxford: Oxford University Press. (A book describing the visual abilities of cortically blind patients.)

Yarbus, A. L. 1967. *Movements and Vision* (trans. B. Haigh). New York: Plenum Press. (A summary of Yarbus's classic experiments on eye movements.)

Key and Classic Papers

Acland, S. M., Aguirre, G. D., Ray, J., et al. 2001. Gene therapy restores vision in a canine model of childhood blindness. *Nat. Gen., 28,* 92–95.

Arden, G. B. 2001. The absence of diabetic retinopathy in patients with retinitis pigmentosa: Implications for pathophysiology and possible treatments. *Br. J. Ophthalmol., 85,* 366–370.

Barlow, H. 1953. Summation and inhibition in the frog's retina. *J. Physiol., 119,* 69–88.

Baylor, D. A. 1987. Photoreceptor signals and vision. *Invest. Ophthalmol. Vis. Sci., 28,* 34–49.

Birol, G., Wang, S., Budzynski, E., Wangsa-Wirawan, N. D., & Linsenmeier, R. A. 2007. Oxygen distribution and consumption in the macaque retina. *Am. J. Physiol. Heart Circ. Physiol., 293*(3), H1696–1704.

Boycott, B. B., & Dowling, J. E. 1969. Organization of the primate retina: light microscopy. *Phil. Trans. R. Soc. Lond. B, 255,* 109–184.

Brown, P. K., & Wald, G. 1964. Visual pigments in single rods and cones of the human retina. *Science, 144,* 45–51.

Dacey, D. M., & Lee, B. B. 1994. The "blue-on" opponent pathway in primate retina originates from a distinct bistratified ganglion cell type. *Nature, 367,* 731–735.

Dowling, J. E. 1960. The chemistry of visual adaptation in the rat. *Nature, 188,* 114–118.

Dowling, J. E., & Boycott, B. B. 1966. Organization of the primate retina: Electron microscopy. *Proc. R. Soc. Lond. B, 166,* 80–111.

Dryja, T. P., McGee, T. L., Reichel, E., et al. 1990. A point mutation of the rhodopsin gene in one form of retinitis pigmentosa. *Nature, 343*(6256), 364–366.

Famiglietti, E. V., Jr., & Kolb, H. 1976. Structural basis for on- and off-center responses in retinal ganglion cells. *Science, 194,* 193–195.

Gross, C. G., Rocha-Miranda, C. E., & Bender, D. B. 1972. Visual properties of neurons in infero-temporal cortex of the macaque. *J. Neurophysiol., 35,* 96–11.

Hartline, H. K. 1938. The response of single optic nerve fibers of the vertebrate eye to illumination of the retina. *Am. J. Physiol., 121,* 400–415.

Hecht, S., Shlaer, S., & Pirenne, M.H. 1942. Energy, quanta, and vision. *J. Gen. Physiol., 25,* 819–840.

Hubel, D. H., & Wiesel, T. N. 1968. Receptive fields and functional architecture of monkey striate cortex. *J. Physiol., 195,* 215–243.

Jacobs, G. H., & Rowe, M. P. 2004. Evolution of vertebrate colour vision. *Clin. Exp. Optom., 87,* 206–216.

Kaneko, A. 1970. Physiological and morphological identification of horizontal, bipolar, and amacrine cells in the goldfish retina. *J. Physiol., 207,* 623–633.

Kaneko, A., & Tachibana, M. 1983. Double color-opponent receptive fields of carp bipolar cells. *Vision Res., 23,* 381–388.

Kuffler, S. W. 1953. Discharge patterns and functional organization of mammalian retina. *J. Neurophysiol., 16,* 37–68.

LaVail, M. M., Yasumura, D., Matthes, M. T., et al. 1998. Protection of mouse photoreceptors by survival factors in retinal degenerations. *Invest. Ophthalmol. Vis. Sci., 39,* 592–602.

Livingstone, M. S., & Hubel, D. H. 1984. Anatomy and physiology of a color system in the primate visual cortex. *J. Neurosci., 4,* 309–350.

MacNichol, E. F., & Svaetichin, G. 1958. Electric responses from the isolated retinas of fishes. *Am. J. Ophthalmol., 46,* 26–46.

Miller, W. H., & Nicol, G. D. 1979. Evidence that cyclic GMP regulates membrane potential of rod photoreceptors. *Nature, 280,* 64–66.

Nathans, J., Piantanida, T. P., Eddy, R. L., Shows, R. L., & Hogness, D. S. 1986. Molecular genetics of inherited variation in human color vision. *Science, 232,* 203–210.

Nathans, J., Thomas, D., & Hogness, D. S. 1986. Molecular genetics of human color vision: The genes encoding blue, green, and red pigments. *Science, 232,* 193–202.

Riggs, L. A., & Ratliff, F. 1953. The effects of counteracting the normal movements of the eye. *J. Opt. Soc. Am., 42,* 872–873.

Roorda, A., & Williams, D. R. 1999. The arrangement of the three cone classes in the living human eye. *Nature, 397,* 520–522.

Roska, B., & Werblin, F. 2001. Vertical interactions across ten parallel, stacked representations in the mammalian retina. *Nature, 410,* 583–587.

Rushton, W. A. H. 1961. Rhodopsin measurement and dark-adaptation in a subject deficient in cone vision. *J. Physiol., 156,* 193–205.

Schiller, P. H., Sandell, H. H., & Maunsell, J. H. R. 1986. Functions of the ON and OFF channels of the visual system. *Nature, 322,* 824–825.

Sjöstrand, F. S. 1953. Ultrastructure of the outer segments of rods and cones of the eye as revealed by the electron microscope. *Cell. Comp. Physiol., 42,* 15–44.

Stryer, L. 1986. Cyclic GMP cascade of vision. *Annu. Rev. Neurosci., 9,* 87–119.

Tomita, T. 1963. Electrical activity in the vertebrate retina. *J. Opt. Soc. Am., 53,* 49–57.

Verhoff, F. H. 1931. Microscopic observations in a case of retinitis pigmentosa. *Arch. Ophthalmol., 5,* 392–407.

Wald, G. 1933. Vitamin A in the retina. *Nature, 132,* 316–317.

Wald, G. 1968. The molecular basis of visual excitation (Nobel Prize lecture). *Nature, 219,* 800–807.

Werblin, F. S., & Dowling, J. E. 1969. Organization of the retina of the mudpuppy, *Necturus maculosus.* II. Intracellular recording. *J. Neurophysiol., 32,* 339–355.

Wurtz, R. H., Goldberg, M. E., & Robinson, D. L. 1982. Brain mechanisms of visual attention. *Sci. Am., 246,* 124–135.

Young, R. W. 1967. The renewal of photoreceptor cell outer segments. *J. Cell Biol., 33,* 61–72.

Young, T. 1802. On the theory of light and colours. *Phil. Trans. R. Soc. Lond., 92,* 12–48.

Index